1 2 3 4

10 9 8 7 6 5

This Book Belongs to :

....

....

Contents

Addition (0-10)

(1) 5 + 1	(2) 3 + 4	(3) 2 + 1	(4) 1 + 4	(5) 7 + 0
(6) 7 + 2	(7) 6 + 1	(8) 5 + 3	(9) 0 + 2	(10) 1 + 3
(11) 1 + 1	(12) 2 + 1	(13) 4 + 2	(14) 0 + 5	(15) 1 + 7
(16) 2 + 3	(17) 4 + 4	(18) 7 + 2	(19) 2 + 6	(20) 4 + 6
(21) 4 + 1	(22) 2 + 0	(23) 9 + 0	(24) 1 + 7	(25) 7 + 3
(26) 3 + 2	(27) 4 + 6	(28) 4 + 2	(29) 5 + 5	(30) 2 + 4

Date:

Name:

Time:

Score: _____ / 30

Addition (0-10)

(1) $\begin{array}{r} 1 \\ +\ 7 \\ \hline \end{array}$.	(2) $\begin{array}{r} 3 \\ +\ 5 \\ \hline \end{array}$.	(3) $\begin{array}{r} 4 \\ +\ 3 \\ \hline \end{array}$.	(4) $\begin{array}{r} 8 \\ +\ 4 \\ \hline \end{array}$ . .	(5) $\begin{array}{r} 5 \\ +\ 2 \\ \hline \end{array}$.
(6) $\begin{array}{r} 6 \\ +\ 1 \\ \hline \end{array}$.	(7) $\begin{array}{r} 2 \\ +\ 7 \\ \hline \end{array}$.	(8) $\begin{array}{r} 5 \\ +\ 6 \\ \hline \end{array}$ . .	(9) $\begin{array}{r} 1\ 0 \\ +\ 0 \\ \hline \end{array}$ . .	(10) $\begin{array}{r} 9 \\ +\ 5 \\ \hline \end{array}$ . .
(11) $\begin{array}{r} 6 \\ +\ 2 \\ \hline \end{array}$.	(12) $\begin{array}{r} 3 \\ +\ 8 \\ \hline \end{array}$ . .	(13) $\begin{array}{r} 6 \\ +\ 6 \\ \hline \end{array}$ . .	(14) $\begin{array}{r} 8 \\ +\ 2 \\ \hline \end{array}$ . .	(15) $\begin{array}{r} 3 \\ +\ 4 \\ \hline \end{array}$.
(16) $\begin{array}{r} 3 \\ +\ 9 \\ \hline \end{array}$ . .	(17) $\begin{array}{r} 6 \\ +\ 6 \\ \hline \end{array}$ . .	(18) $\begin{array}{r} 5 \\ +\ 7 \\ \hline \end{array}$ . .	(19) $\begin{array}{r} 9 \\ +\ 2 \\ \hline \end{array}$ . .	(20) $\begin{array}{r} 6 \\ +\ 3 \\ \hline \end{array}$.
(21) $\begin{array}{r} 9 \\ +\ 3 \\ \hline \end{array}$ . .	(22) $\begin{array}{r} 5 \\ +\ 9 \\ \hline \end{array}$ . .	(23) $\begin{array}{r} 8 \\ +\ 1 \\ \hline \end{array}$.	(24) $\begin{array}{r} 4 \\ +\ 9 \\ \hline \end{array}$ . .	(25) $\begin{array}{r} 8 \\ +\ 4 \\ \hline \end{array}$ . .
(26) $\begin{array}{r} 7 \\ +\ 7 \\ \hline \end{array}$ . .	(27) $\begin{array}{r} 5 \\ +\ 3 \\ \hline \end{array}$.	(28) $\begin{array}{r} 2 \\ +\ 9 \\ \hline \end{array}$ . .	(29) $\begin{array}{r} 1 \\ +\ 0 \\ \hline \end{array}$.	(30) $\begin{array}{r} 5 \\ +\ 2 \\ \hline \end{array}$.

Date:

Time:

Name:

Score: ___ / 30

Addition (0-10)

1	2	3	4	5
9 + 6 . .	8 + 7 . .	3 + 9 . .	8 + 5 . .	6 + 6 . .
6 6 + 7 . .	**7** 8 + 7 . .	**8** 1 0 + 4 . .	**9** 8 + 1 0 . .	**10** 9 + 6 . .
11 2 + 7 .	**12** 1 0 + 6 . .	**13** 9 + 3 . .	**14** 4 + 6 . .	**15** 9 + 8 . .
16 7 + 2 .	**17** 5 + 5 . .	**18** 6 + 7 . .	**19** 0 + 9 .	**20** 1 0 + 1 . .
21 7 + 8 . .	**22** 1 + 1 0 . .	**23** 3 + 1 .	**24** 9 + 5 . .	**25** 7 + 3 . .
26 9 + 9 . .	**27** 2 + 9 . .	**28** 4 + 6 . .	**29** 1 0 + 7 . .	**30** 8 + 8 . .

Date:

Name:

Time:

Score: _____ / 30

Addition (0-10)

(1) 7 + 5	(2) 5 + 9	(3) 4 + 7

1.
```
   7
+  5
. .
```

2.
```
   5
+  9
. .
```

3.
```
   4
+  7
. .
```

4.
```
   2
+  8
. .
```

5.
```
   9
+  6
. .
```

6.
```
   8
+  8
. .
```

7.
```
  1 0
+   7
. .
```

8.
```
   9
+ 1 0
. .
```

9.
```
   3
+  9
. .
```

10.
```
   9
+  3
. .
```

11.
```
   3
+  6
. .
```

12.
```
   5
+  8
. .
```

13.
```
   1
+  9
. .
```

14.
```
   8
+  7
. .
```

15.
```
   6
+  6
. .
```

16.
```
   5
+  9
. .
```

17.
```
  1 0
+   5
. .
```

18.
```
  1 0
+   8
. .
```

19.
```
   6
+  3
. .
```

20.
```
   8
+  6
. .
```

21.
```
   2
+  9
. .
```

22.
```
   8
+  3
. .
```

23.
```
   7
+  8
. .
```

24.
```
   3
+  8
. .
```

25.
```
   8
+  4
. .
```

26.
```
   2
+  9
. .
```

27.
```
   5
+  6
. .
```

28.
```
   6
+  9
. .
```

29.
```
   5
+  8
. .
```

30.
```
  1 0
+   1
. .
```

Date: ....

Time: ....

Name: ....

Score: .... / 30

Addition (0-10)

1) 1 0 + 1 0	2) 8 + 4	3) 6 + 9	4) 9 + 6	5) 8 + 9
6) 7 + 8	7) 9 + 9	8) 1 0 + 2	9) 8 + 5	10) 6 + 8
11) 8 + 3	12) 9 + 4	13) 1 0 + 8	14) 2 + 1 0	15) 8 + 8
16) 8 + 8	17) 4 + 9	18) 5 + 7	19) 5 + 6	20) 7 + 7
21) 1 0 + 0	22) 5 + 8	23) 7 + 3	24) 7 + 9	25) 8 + 7
26) 4 + 9	27) 7 + 5	28) 6 + 7	29) 3 + 8	30) 1 0 + 3

Date: .... Time: ....

Name: .... Score: .... / 30

Addition (10-20)

(1) $\begin{array}{r} 1\ 1 \\ +\ 1\ 0 \\ \hline \end{array}$	(2) $\begin{array}{r} 1\ 2 \\ +\ 1\ 3 \\ \hline \end{array}$	(3) $\begin{array}{r} 1\ 5 \\ +\ 1\ 2 \\ \hline \end{array}$

(1) 11 + 10

(2) 12 + 13

(3) 15 + 12

(4) 13 + 16

(5) 11 + 18

(6) 17 + 12

(7) 10 + 19

(8) 14 + 12

(9) 12 + 15

(10) 15 + 12

(11) 11 + 13

(12) 10 + 14

(13) 13 + 15

(14) 15 + 10

(15) 10 + 10

(16) 17 + 12

(17) 14 + 13

(18) 18 + 10

(19) 15 + 13

(20) 16 + 11

(21) 15 + 10

(22) 16 + 12

(23) 17 + 11

(24) 10 + 18

(25) 12 + 17

(26) 14 + 14

(27) 15 + 12

(28) 16 + 10

(29) 13 + 15

(30) 10 + 15

Date: ....   Time: ....

Name: ....   Score: .... / 30

Addition (10-20)

① 13 + 16 . .	② 11 + 11 . .	③ 12 + 12 . .

1) 13
  + 16
   . .

2) 11
  + 11
   . .

3) 12
  + 12
   . .

4) 15
  + 10
   . .

5) 16
  + 12
   . .

6) 11
  + 17
   . .

7) 14
  + 13
   . .

8) 15
  + 11
   . .

9) 10
  + 17
   . .

10) 17
  + 12
   . .

11) 18
  + 10
   . .

12) 13
  + 17
   . .

13) 19
  + 12
   . .

14) 11
  + 17
   . .

15) 14
  + 15
   . .

16) 18
  + 13
   . .

17) 16
  + 10
   . .

18) 14
  + 14
   . .

19) 19
  + 14
   . .

20) 11
  + 15
   . .

21) 13
  + 12
   . .

22) 17
  + 17
   . .

23) 14
  + 10
   . .

24) 17
  + 12
   . .

25) 12
  + 18
   . .

26) 12
  + 19
   . .

27) 13
  + 15
   . .

28) 16
  + 11
   . .

29) 14
  + 17
   . .

30) 17
  + 10
   . .

Date:

Name:

Time:

Score: _____ / 30

Addition (10-20)

1	2	3	4	5
2 0 + 1 3	1 5 + 1 8	1 4 + 1 6	1 9 + 1 2	1 2 + 1 4

6	7	8	9	10
1 8 + 1 5	1 7 + 1 4	1 1 + 1 0	1 3 + 1 3	1 0 + 1 8

11	12	13	14	15
2 0 + 1 0	1 8 + 1 1	1 2 + 1 8	1 1 + 1 4	1 3 + 1 2

16	17	18	19	20
1 5 + 1 7	1 3 + 1 3	1 8 + 1 0	1 2 + 2 0	1 0 + 1 7

21	22	23	24	25
2 0 + 1 2	1 2 + 1 4	1 9 + 1 7	1 3 + 1 5	1 5 + 1 0

26	27	28	29	30
1 6 + 1 7	1 5 + 1 3	1 8 + 1 8	1 7 + 1 4	1 9 + 1 0

Date: ....

Name: ....

Time: ....

Score: .... / 30

Addition (10-20)

①
```
  1 0
+ 1 9
-----
  . .
```

②
```
  2 0
+ 1 6
-----
  . .
```

③
```
  1 7
+ 1 8
-----
  . .
```

④
```
  1 0
+ 1 6
-----
  . .
```

⑤
```
  1 8
+ 1 3
-----
  . .
```

⑥
```
  1 6
+ 1 9
-----
  . .
```

⑦
```
  1 5
+ 1 3
-----
  . .
```

⑧
```
  1 3
+ 2 0
-----
  . .
```

⑨
```
  1 7
+ 1 4
-----
  . .
```

⑩
```
  2 0
+ 1 8
-----
  . .
```

⑪
```
  1 5
+ 1 4
-----
  . .
```

⑫
```
  1 9
+ 1 0
-----
  . .
```

⑬
```
  1 5
+ 2 0
-----
  . .
```

⑭
```
  1 4
+ 1 3
-----
  . .
```

⑮
```
  1 5
+ 1 6
-----
  . .
```

⑯
```
  1 7
+ 2 0
-----
  . .
```

⑰
```
  1 6
+ 1 8
-----
  . .
```

⑱
```
  1 9
+ 1 1
-----
  . .
```

⑲
```
  2 0
+ 2 0
-----
  . .
```

⑳
```
  1 8
+ 1 0
-----
  . .
```

㉑
```
  1 8
+ 1 5
-----
  . .
```

㉒
```
  1 7
+ 1 3
-----
  . .
```

㉓
```
  1 7
+ 2 0
-----
  . .
```

㉔
```
  1 4
+ 1 8
-----
  . .
```

㉕
```
  1 3
+ 1 6
-----
  . .
```

㉖
```
  1 4
+ 1 4
-----
  . .
```

㉗
```
  1 6
+ 1 7
-----
  . .
```

㉘
```
  1 5
+ 1 7
-----
  . .
```

㉙
```
  1 5
+ 2 0
-----
  . .
```

㉚
```
  1 9
+ 1 1
-----
  . .
```

Date:

Name:

Time:

Score: _____ / 30

Addition (10-20)

1)
```
   1 2
 + 2 0
 ------
```

2)
```
   2 0
 + 1 3
 ------
```

3)
```
   1 4
 + 1 8
 ------
```

4)
```
   2 0
 + 1 7
 ------
```

5)
```
   1 9
 + 1 2
 ------
```

6)
```
   1 4
 + 1 9
 ------
```

7)
```
   1 0
 + 1 7
 ------
```

8)
```
   1 6
 + 1 4
 ------
```

9)
```
   1 8
 + 1 5
 ------
```

10)
```
   2 0
 + 1 3
 ------
```

11)
```
   1 2
 + 1 7
 ------
```

12)
```
   1 5
 + 1 6
 ------
```

13)
```
   1 8
 + 2 0
 ------
```

14)
```
   2 0
 + 2 0
 ------
```

15)
```
   1 5
 + 1 7
 ------
```

16)
```
   1 1
 + 1 0
 ------
```

17)
```
   1 6
 + 2 0
 ------
```

18)
```
   1 9
 + 1 9
 ------
```

19)
```
   2 0
 + 1 5
 ------
```

20)
```
   1 7
 + 1 0
 ------
```

21)
```
   1 8
 + 1 4
 ------
```

22)
```
   2 0
 + 1 3
 ------
```

23)
```
   1 2
 + 1 0
 ------
```

24)
```
   1 5
 + 1 8
 ------
```

25)
```
   1 4
 + 1 6
 ------
```

26)
```
   1 8
 + 1 2
 ------
```

27)
```
   2 0
 + 1 5
 ------
```

28)
```
   1 5
 + 1 5
 ------
```

29)
```
   1 5
 + 1 9
 ------
```

30)
```
   2 0
 + 1 0
 ------
```

Date: ....

Name: ....

Time: ....

Score: .... / 30

Subtraction (0-10)

1) 1 − 0 .	2) 2 − 0 .	3) 2 − 1 .	4) 3 − 1 .	5) 3 − 2 .
6) 4 − 2 .	7) 4 − 3 .	8) 5 − 3 .	9) 5 − 5 .	10) 6 − 4 .
11) 5 − 4 .	12) 6 − 5 .	13) 7 − 4 .	14) 8 − 3 .	15) 7 − 3 .
16) 7 − 6 .	17) 8 − 6 .	18) 9 − 7 .	19) 9 − 5 . .	20) 4 − 0 .
21) 8 − 4 .	22) 4 − 3 .	23) 2 − 0 .	24) 6 − 3 .	25) 6 − 6 .
26) 8 − 2 .	27) 6 − 5 .	28) 9 − 8 .	29) 9 − 4 .	30) 1 0 − 9 .

Date: ....

Name: ....

Time: ....

Score: .... / 30

Subtraction (0-10)

① 4 − 1 .	② 5 − 2 .	③ 6 − 2 .	④ 7 − 3 .	⑤ 8 − 5 .
⑥ 9 − 3 .	⑦ 1 0 − 5 .	⑧ 5 − 1 .	⑨ 7 − 2 .	⑩ 9 − 2 .
⑪ 8 − 2 .	⑫ 8 − 4 .	⑬ 9 − 8 .	⑭ 1 0 − 1 .	⑮ 8 − 8 .
⑯ 1 0 − 2 .	⑰ 9 − 1 .	⑱ 7 − 1 .	⑲ 6 − 3 .	⑳ 4 − 4 .
㉑ 8 − 5 .	㉒ 9 − 6 .	㉓ 1 0 − 6 .	㉔ 9 − 5 .	㉕ 8 − 3 .
㉖ 9 − 1 .	㉗ 1 0 − 9 .	㉘ 8 − 7 .	㉙ 7 − 2 .	㉚ 9 − 7 .

Date: ....   Time: ....

Name: ....   Score: .... / 30

Subtraction (0-10)

(1) 5 − 2 .	(2) 8 − 3 .	(3) 6 − 4 .	(4) 9 − 3 .	(5) 7 − 1 .
(6) 1 0 − 3 .	(7) 7 − 3 .	(8) 9 − 4 .	(9) 6 − 2 .	(10) 8 − 1 .
(11) 1 0 − 7 .	(12) 5 − 3 .	(13) 9 − 6 .	(14) 7 − 2 .	(15) 1 0 − 3 .
(16) 8 − 2 .	(17) 3 − 1 .	(18) 9 − 5 .	(19) 7 − 3 .	(20) 1 0 − 4 .
(21) 7 − 4 .	(22) 2 − 1 .	(23) 8 − 4 .	(24) 6 − 1 .	(25) 9 − 5 .
(26) 7 − 2 .	(27) 4 − 1 .	(28) 3 − 2 .	(29) 1 0 − 8 .	(30) 6 − 5 .

Subtraction (0-10)

(1) 9 − 7 .	(2) 7 − 5 .	(3) 8 − 6 .	(4) 4 − 2 .	(5) 1 0 − 9 .
(6) 9 − 1 .	(7) 6 − 4 .	(8) 7 − 3 .	(9) 5 − 2 .	(10) 8 − 5 .
(11) 1 0 − 6 .	(12) 7 − 1 .	(13) 9 − 3 .	(14) 6 − 2 .	(15) 8 − 1 .
(16) 1 0 − 7 .	(17) 5 − 3 .	(18) 9 − 6 .	(19) 7 − 2 .	(20) 1 0 − 3 .
(21) 8 − 2 .	(22) 3 − 1 .	(23) 9 − 5 .	(24) 6 − 3 .	(25) 1 0 − 4 .
(26) 7 − 4 .	(27) 2 − 1 .	(28) 8 − 4 .	(29) 6 − 1 .	(30) 9 − 5 .

Date: ....

Time: ....

Name: ....

Score: .... / 30

Subtraction (0-10)

(1) 9 − 2 .	(2) 8 − 3 .	(3) 1 0 − 4 .	(4) 7 − 1 .	(5) 6 − 4 .
(6) 1 0 − 6 .	(7) 7 − 3 .	(8) 9 − 4 .	(9) 8 − 1 .	(10) 1 0 − 7 .
(11) 9 − 6 .	(12) 7 − 2 .	(13) 1 0 − 3 .	(14) 8 − 2 .	(15) 9 − 1 .
(16) 6 − 3 .	(17) 1 0 − 4 .	(18) 7 − 4 .	(19) 8 − 6 .	(20) 1 0 − 9 .
(21) 9 − 1 .	(22) 6 − 5 .	(23) 9 − 7 .	(24) 7 − 5 .	(25) 8 − 4 .
(26) 1 0 − 6 .	(27) 9 − 3 .	(28) 8 − 5 .	(29) 7 − 2 .	(30) 1 0 − 3 .

Subtraction (10-20)

1	2	3	4	5
1 5 - 1 0 . .	1 8 - 1 1 . .	1 4 - 1 0 . .	1 7 - 1 3 . .	1 9 - 1 1 . .

6	7	8	9	10
2 0 - 1 2 . .	1 6 - 1 0 . .	2 0 - 1 3 . .	1 2 - 1 0 . .	1 9 - 1 4 . .

11	12	13	14	15
1 3 - 1 0 . .	1 6 - 1 2 . .	1 8 - 1 3 . .	2 0 - 1 5 . .	1 7 - 1 1 . .

16	17	18	19	20
1 9 - 1 6 . .	1 4 - 1 2 . .	2 0 - 1 7 . .	1 1 - 1 0 . .	1 8 - 1 5 . .

21	22	23	24	25
1 5 - 1 2 . .	2 0 - 1 8 . .	1 6 - 1 4 . .	1 9 - 1 7 . .	1 3 - 1 2 . .

26	27	28	29	30
2 0 - 1 9 . .	1 7 - 1 5 . .	1 8 - 1 6 . .	1 4 - 1 3 . .	2 0 - 2 0 . .

Date: ....

Time: ....

Name: ....

Score: .... / 30

Subtraction (10-20)

1)
```
   1 8
-  1 1
```

2)
```
   1 9
-  1 3
```

3)
```
   1 7
-  1 2
```

4)
```
   2 0
-  1 5
```

5)
```
   1 6
-  1 1
```

6)
```
   1 4
-  1 0
```

7)
```
   2 0
-  1 6
```

8)
```
   1 9
-  1 4
```

9)
```
   1 8
-  1 2
```

10)
```
   1 7
-  1 0
```

11)
```
   1 6
-  1 4
```

12)
```
   1 3
-  1 0
```

13)
```
   2 0
-  1 7
```

14)
```
   1 5
-  1 2
```

15)
```
   1 9
-  1 5
```

16)
```
   1 8
-  1 3
```

17)
```
   1 7
-  1 1
```

18)
```
   1 6
-  1 2
```

19)
```
   2 0
-  1 8
```

20)
```
   1 4
-  1 3
```

21)
```
   1 9
-  1 6
```

22)
```
   1 5
-  1 0
```

23)
```
   1 8
-  1 4
```

24)
```
   1 7
-  1 5
```

25)
```
   2 0
-  1 9
```

26)
```
   1 6
-  1 3
```

27)
```
   1 9
-  1 7
```

28)
```
   1 8
-  1 5
```

29)
```
   1 7
-  1 3
```

30)
```
   1 6
-  1 0
```

Date: ....

Name: ....

Time: ....

Score: .... / 30

Subtraction (10-20)

1 20 − 20	2 19 − 18	3 18 − 16	4 17 − 14	5 16 − 15
6 15 − 14	7 14 − 12	8 13 − 10	9 20 − 11	10 19 − 12
11 18 − 17	12 17 − 16	13 15 − 13	14 14 − 11	15 16 − 14
16 13 − 10	17 20 − 13	18 19 − 15	19 18 − 10	20 17 − 12
21 16 − 11	22 16 − 16	23 14 − 10	24 20 − 14	25 19 − 16
26 18 − 13	27 17 − 15	28 16 − 12	29 15 − 11	30 14 − 12

Date: ....

Time: ....

Name: ....

Score: .... / 30

Subtraction (10-20)

① 19 - 10 . .	② 20 - 11 . .	③ 18 - 12 . .	④ 17 - 13 . .	⑤ 16 - 10 . .
⑥ 20 - 12 . .	⑦ 20 - 12 . .	⑧ 19 - 13 . .	⑨ 18 - 17 . .	⑩ 19 - 10 . .
⑪ 16 - 11 . .	⑫ 20 - 14 . .	⑬ 19 - 15 . .	⑭ 18 - 10 . .	⑮ 17 - 12 . .
⑯ 16 - 13 . .	⑰ 15 - 14 . .	⑱ 20 - 16 . .	⑲ 19 - 17 . .	⑳ 18 - 16 . .
㉑ 16 - 15 . .	㉒ 15 - 12 . .	㉓ 20 - 17 . .	㉔ 19 - 18 . .	㉕ 19 - 16 . .
㉖ 17 - 15 . .	㉗ 16 - 14 . .	㉘ 15 - 13 . .	㉙ 20 - 19 . .	㉚ 18 - 11 . .

Date: ....

Name: ....

Time: ....

Score: .... / 30

Subtraction (10-20)

(1) 18 − 11	(2) 17 − 10	(3) 16 − 12	(4) 20 − 13	(5) 19 − 17
(6) 18 − 15	(7) 17 − 12	(8) 16 − 15	(9) 15 − 11	(10) 20 − 15
(11) 19 − 16	(12) 18 − 14	(13) 17 − 12	(14) 16 − 10	(15) 15 − 10
(16) 20 − 16	(17) 19 − 17	(18) 18 − 13	(19) 17 − 14	(20) 16 − 11
(21) 15 − 12	(22) 14 − 13	(23) 20 − 18	(24) 19 − 15	(25) 18 − 17
(26) 11 − 10	(27) 16 − 13	(28) 15 − 11	(29) 14 − 12	(30) 13 − 10

Date: ....

Time: ....

Name: ....

Score: .... / 30

Addition & Subtraction (0-20)

1	2	3	4	5
5 + 1 2 . .	2 0 - 3 . .	8 + 9 . .	1 4 - 2 . .	1 0 + 6 . .
6	**7**	**8**	**9**	**10**
1 8 - 8 . .	3 + 1 1 . .	2 0 - 5 . .	2 + 1 3 . .	1 6 - 6 . .
11	**12**	**13**	**14**	**15**
7 + 9 . .	1 5 - 3 . .	4 + 1 0 . .	1 3 - 4 . .	9 + 8 . .
16	**17**	**18**	**19**	**20**
2 0 - 2 . .	1 + 1 4 . .	1 7 - 6 . .	6 + 1 1 . .	2 0 - 7 . .
21	**22**	**23**	**24**	**25**
3 + 1 2 . .	1 2 - 4 . .	8 + 1 1 . .	2 0 - 8 . .	2 + 1 5 . .
26	**27**	**28**	**29**	**30**
1 8 - 5 . .	5 + 1 3 . .	1 4 - 5 . .	1 0 + 7 . .	1 5 - 4 . .

Date: ....

Name: ....

Time: ....

Score: .... / 30

Addition & Subtraction (0-20)

① 4 + 1 1 — — . .	② 1 3 − 3 — — . .	③ 1 0 + 9 — — . .	④ 2 0 − 6 — — . .	⑤ 1 6 + 1 — — . .
⑥ 1 7 − 4 — — . .	⑦ 1 2 + 7 — — . .	⑧ 1 6 − 2 — — . .	⑨ 6 + 9 — — . .	⑩ 2 0 − 1 0 — — . .
⑪ 1 3 + 3 — — . .	⑫ 1 4 − 4 — — . .	⑬ 1 1 + 7 — — . .	⑭ 1 8 − 3 — — . .	⑮ 8 + 1 0 — — . .
⑯ 2 0 − 5 — — . .	⑰ 2 + 1 4 — — . .	⑱ 1 6 − 3 — — . .	⑲ 1 2 + 7 — — . .	⑳ 1 5 − 2 — — . .
㉑ 1 0 + 6 — — . .	㉒ 1 9 − 5 — — . .	㉓ 4 + 1 2 — — . .	㉔ 1 3 − 2 — — . .	㉕ 9 + 7 — — . .
㉖ 1 6 − 1 — — . .	㉗ 1 5 + 1 — — . .	㉘ 1 9 − 2 — — . .	㉙ 7 + 1 1 — — . .	㉚ 2 0 − 1 — — . .

Date: ....   Time: ....

Name: ....   Score: .... / 30

Addition & Subtraction (0-20)

1)
```
    3
+ 1 0
-----
  . .
```

2)
```
  1 4
-   3
-----
  . .
```

3)
```
    8
+   9
-----
  . .
```

4)
```
  1 9
-   4
-----
  . .
```

5)
```
    6
+   8
-----
  . .
```

6)
```
  1 5
-   1
-----
  . .
```

7)
```
    5
+   9
-----
  . .
```

8)
```
  1 8
-   3
-----
  . .
```

9)
```
    4
+   8
-----
  . .
```

10)
```
  2 0
-   1
-----
  . .
```

11)
```
  1 2
+   2
-----
  . .
```

12)
```
  1 7
-   1
-----
  . .
```

13)
```
  1 0
+   7
-----
  . .
```

14)
```
  1 6
-   4
-----
  . .
```

15)
```
    9
+   6
-----
  . .
```

16)
```
  1 4
-   1
-----
  . .
```

17)
```
    7
+   6
-----
  . .
```

18)
```
  2 0
-   3
-----
  . .
```

19)
```
  1 1
+   3
-----
  . .
```

20)
```
  1 3
-   1
-----
  . .
```

21)
```
    5
+   6
-----
  . .
```

22)
```
  1 8
-   4
-----
  . .
```

23)
```
    8
+   7
-----
  . .
```

24)
```
  1 5
-   2
-----
  . .
```

25)
```
    4
+   5
-----
    .
```

26)
```
  1 3
-   1
-----
  . .
```

27)
```
    6
+   4
-----
  . .
```

28)
```
  2 0
-   4
-----
  . .
```

29)
```
  1 1
+   2
-----
  . .
```

30)
```
  1 2
-   1
-----
  . .
```

Date: .... Time: ....

Name: .... Score: .... / 30

Addition & Subtraction (0-20)

① 1 2 + 4 ——— . .	② 1 7 − 7 ——— . .	③ 7 + 8 ——— . .	④ 1 4 − 6 ——— . .	⑤ 1 0 + 7 ——— . .
⑥ 1 9 − 9 ——— . .	⑦ 1 0 + 3 ——— . .	⑧ 1 7 − 5 ——— . .	⑨ 1 4 + 2 ——— . .	⑩ 1 6 − 5 ——— . .
⑪ 9 + 7 ——— . .	⑫ 1 5 − 3 ——— . .	⑬ 1 1 + 4 ——— . .	⑭ 1 3 − 4 ——— . .	⑮ 9 + 9 ——— . .
⑯ 2 0 − 2 ——— . .	⑰ 1 5 + 1 ——— . .	⑱ 1 7 − 6 ——— . .	⑲ 1 2 + 6 ——— . .	⑳ 2 0 − 8 ——— . .
㉑ 1 3 + 3 ——— . .	㉒ 1 2 − 3 ——— . .	㉓ 1 0 + 8 ——— . .	㉔ 1 9 − 5 ——— . .	㉕ 1 6 + 2 ——— .
㉖ 1 8 − 4 ——— . .	㉗ 1 2 + 5 ——— . .	㉘ 1 4 − 4 ——— . .	㉙ 1 0 + 8 ——— . .	㉚ 1 5 − 2 ——— . .

Date: ....

Time: ....

Name: ....

Score: .... / 30

Addition & Subtraction (0-20)

(1) $\begin{array}{r} 1\ 0 \\ +\ \ 5 \\ \hline \end{array}$	(2) $\begin{array}{r} 1\ 3 \\ -\ \ 3 \\ \hline \end{array}$	(3) $\begin{array}{r} 1\ 1 \\ +\ \ 9 \\ \hline \end{array}$	(4) $\begin{array}{r} 2\ 0 \\ -\ \ 7 \\ \hline \end{array}$	(5) $\begin{array}{r} 1\ 6 \\ +\ \ 1 \\ \hline \end{array}$
(6) $\begin{array}{r} 1\ 7 \\ -\ \ 4 \\ \hline \end{array}$	(7) $\begin{array}{r} 1\ 3 \\ +\ \ 7 \\ \hline \end{array}$	(8) $\begin{array}{r} 1\ 6 \\ -\ \ 2 \\ \hline \end{array}$	(9) $\begin{array}{r} 1\ 0 \\ +\ \ 6 \\ \hline \end{array}$	(10) $\begin{array}{r} 2\ 0 \\ -\ 1\ 0 \\ \hline \end{array}$
(11) $\begin{array}{r} 1\ 2 \\ +\ \ 3 \\ \hline \end{array}$	(12) $\begin{array}{r} 1\ 4 \\ -\ \ 5 \\ \hline \end{array}$	(13) $\begin{array}{r} 1\ 1 \\ +\ \ 7 \\ \hline \end{array}$	(14) $\begin{array}{r} 1\ 8 \\ -\ \ 3 \\ \hline \end{array}$	(15) $\begin{array}{r} 1\ 1 \\ +\ \ 8 \\ \hline \end{array}$
(16) $\begin{array}{r} 2\ 0 \\ -\ \ 6 \\ \hline \end{array}$	(17) $\begin{array}{r} 1\ 5 \\ +\ \ 2 \\ \hline \end{array}$	(18) $\begin{array}{r} 1\ 7 \\ -\ \ 3 \\ \hline \end{array}$	(19) $\begin{array}{r} 1\ 3 \\ +\ \ 5 \\ \hline \end{array}$	(20) $\begin{array}{r} 1\ 5 \\ -\ \ 1 \\ \hline \end{array}$
(21) $\begin{array}{r} 1\ 0 \\ +\ \ 6 \\ \hline \end{array}$	(22) $\begin{array}{r} 1\ 9 \\ -\ \ 5 \\ \hline \end{array}$	(23) $\begin{array}{r} 1\ 2 \\ +\ \ 4 \\ \hline \end{array}$	(24) $\begin{array}{r} 1\ 3 \\ -\ \ 2 \\ \hline \end{array}$	(25) $\begin{array}{r} 9 \\ +\ \ 8 \\ \hline \end{array}$
(26) $\begin{array}{r} 1\ 6 \\ -\ \ 1 \\ \hline \end{array}$	(27) $\begin{array}{r} 1\ 8 \\ +\ \ 2 \\ \hline \end{array}$	(28) $\begin{array}{r} 1\ 7 \\ -\ \ 3 \\ \hline \end{array}$	(29) $\begin{array}{r} 1\ 1 \\ +\ \ 7 \\ \hline \end{array}$	(30) $\begin{array}{r} 2\ 0 \\ -\ \ 5 \\ \hline \end{array}$

Date: ....

Time: ....

Name: ....

Score: .... / 30

Addition & Subtraction (0-20)

①	②	③	④	⑤
1 3 + 5	1 0 - 4	1 2 + 4	9 - 3	4 + 3

⑥	⑦	⑧	⑨	⑩
1 1 - 6	1 1 + 2	7 - 1	6 + 1	1 3 - 8

⑪	⑫	⑬	⑭	⑮
2 + 1	9 - 5	4 + 2	1 2 - 8	3 + 1

⑯	⑰	⑱	⑲	⑳
8 - 2	7 + 3	1 5 - 9	1 + 4	1 0 - 5

㉑	㉒	㉓	㉔	㉕
5 + 2	1 1 - 7	2 + 3	8 - 4	1 6 + 4

㉖	㉗	㉘	㉙	㉚
1 3 - 9	1 2 + 3	1 9 - 5	1 4 + 1	1 2 - 7

Date: .... Time: ....

Name: .... Score: .... / 30

Addition & Subtraction (0-20)

①   1 3
 +   2
    . .
	② 1 7
	 - 2
	    . .
		③ 1 0
		 + 5
		    . .
			④ 1 5
			 - 8
			    . .
				⑤ 1 2
				 + 4
				    . .

⑥   1 2
 -   5
    . .
	⑦ 1 0
	 + 1
	    . .
		⑧ 1 5
		 - 1
		    . .
			⑨ 1 4
			 + 6
			    . .
				⑩ 1 3
				 - 4
				    . .

⑪   2 0
 +   5
    . .
	⑫ 1 8
	 - 3
	    . .
		⑬ 1 7
		 + 2
		    . .
			⑭ 1 4
			 - 6
			    . .
				⑮ 1 4
				 + 3
				    . .

⑯   1 2
 -   3
    . .
	⑰ 1 0
	 + 3
	    . .
		⑱ 1 6
		 - 4
		    . .
			⑲ 1 5
			 + 6
			    . .
				⑳ 1 2
				 - 1 2
				    . .

㉑   1 6
 +   2
    . .
	㉒ 1 3
	 - 5
	    . .
		㉓ 2 0
		 + 1 0
		    . .
			㉔ 1 8
			 - 2
			    . .
				㉕ 1 5
				 + 3
				    . .

㉖   1 2
 -   4
    . .
	㉗ 1 0
	 + 7
	    . .
		㉘ 1 9
		 - 2
		    . .
			㉙ 1 7
			 + 1
			    . .
				㉚ 1 4
				 - 7
				    . .

Date: ....

Name: ....

Time: ....

Score: .... / 30

Addition & Subtraction (0-20)

① $\begin{array}{r} 3 \\ +\ 1\ 3 \\ \hline \end{array}$	② $\begin{array}{r} 1\ 0 \\ -\ \ \ 1 \\ \hline \end{array}$	③ $\begin{array}{r} 1\ 6 \\ +\ \ \ 4 \\ \hline \end{array}$	④ $\begin{array}{r} 1\ 3 \\ -\ \ \ 6 \\ \hline \end{array}$	⑤ $\begin{array}{r} 1\ 1 \\ +\ \ \ 8 \\ \hline \end{array}$
⑥ $\begin{array}{r} 1\ 8 \\ -\ \ \ 1 \\ \hline \end{array}$	⑦ $\begin{array}{r} 1\ 5 \\ +\ \ \ 4 \\ \hline \end{array}$	⑧ $\begin{array}{r} 1\ 1 \\ -\ \ \ 3 \\ \hline \end{array}$	⑨ $\begin{array}{r} 1\ 2 \\ +\ \ \ 8 \\ \hline \end{array}$	⑩ $\begin{array}{r} 1\ 9 \\ -\ \ \ 1 \\ \hline \end{array}$
⑪ $\begin{array}{r} 2\ 0 \\ +\ \ \ 4 \\ \hline \end{array}$	⑫ $\begin{array}{r} 1\ 5 \\ -\ \ \ 7 \\ \hline \end{array}$	⑬ $\begin{array}{r} 1\ 3 \\ +\ \ \ 6 \\ \hline \end{array}$	⑭ $\begin{array}{r} 1\ 2 \\ -\ \ \ 3 \\ \hline \end{array}$	⑮ $\begin{array}{r} 1\ 1 \\ +\ \ \ 9 \\ \hline \end{array}$
⑯ $\begin{array}{r} 1\ 0 \\ -\ 1\ 0 \\ \hline \end{array}$	⑰ $\begin{array}{r} 1\ 4 \\ +\ \ \ 5 \\ \hline \end{array}$	⑱ $\begin{array}{r} 1\ 4 \\ -\ \ \ 5 \\ \hline \end{array}$	⑲ $\begin{array}{r} 1\ 2 \\ +\ \ \ 9 \\ \hline \end{array}$	⑳ $\begin{array}{r} 1\ 8 \\ -\ \ \ 9 \\ \hline \end{array}$
㉑ $\begin{array}{r} 1\ 6 \\ +\ \ \ 9 \\ \hline \end{array}$	㉒ $\begin{array}{r} 2\ 0 \\ -\ 1\ 0 \\ \hline \end{array}$	㉓ $\begin{array}{r} 1\ 6 \\ +\ \ \ 5 \\ \hline \end{array}$	㉔ $\begin{array}{r} 1\ 3 \\ -\ \ \ 4 \\ \hline \end{array}$	㉕ $\begin{array}{r} 1\ 1 \\ +\ 1\ 0 \\ \hline \end{array}$
㉖ $\begin{array}{r} 1\ 9 \\ -\ \ \ 9 \\ \hline \end{array}$	㉗ $\begin{array}{r} 1\ 5 \\ +\ 1\ 3 \\ \hline \end{array}$	㉘ $\begin{array}{r} 2\ 0 \\ -\ \ \ 1 \\ \hline \end{array}$	㉙ $\begin{array}{r} 1\ 2 \\ +\ \ \ 8 \\ \hline \end{array}$	㉚ $\begin{array}{r} 1\ 6 \\ -\ \ \ 5 \\ \hline \end{array}$

Date: ....

Time: ....

Name: ....

Score: .... / 30

Addition & Subtraction (0-20)

1)
```
  1 3
+ 1 5
-----
  . .
```

2)
```
  1 9
-   4
-----
  . .
```

3)
```
  1 3
+   4
-----
  . .
```

4)
```
  1 0
-   6
-----
  . .
```

5)
```
  1 5
+   3
-----
  . .
```

6)
```
  1 2
-   6
-----
  . .
```

7)
```
  1 1
+   6
-----
  . .
```

8)
```
  1 7
-   4
-----
  . .
```

9)
```
  1 6
+   2
-----
  . .
```

10)
```
  1 3
-   8
-----
  . .
```

11)
```
    2
+ 1 1
-----
  . .
```

12)
```
  1 9
-   6
-----
  . .
```

13)
```
  1 4
+   2
-----
  . .
```

14)
```
  1 1
-   8
-----
  . .
```

15)
```
  1 3
+   3
-----
  . .
```

16)
```
  1 8
-   2
-----
  . .
```

17)
```
  2 0
+   5
-----
  . .
```

18)
```
  1 5
-   9
-----
  . .
```

19)
```
  1 1
+   4
-----
  . .
```

20)
```
  1 0
-   5
-----
  . .
```

21)
```
  1 5
+   5
-----
  . .
```

22)
```
  1 2
-   7
-----
  . .
```

23)
```
  1 3
+   2
-----
  . .
```

24)
```
  1 8
-   6
-----
  . .
```

25)
```
  2 0
+   7
-----
  . .
```

26)
```
  1 3
-   6
-----
  . .
```

27)
```
  1 1
+   9
-----
  . .
```

28)
```
  1 9
-   5
-----
  . .
```

29)
```
  1 3
+   8
-----
  . .
```

30)
```
  1 1
-   7
-----
  . .
```

Date: ....

Time: ....

Name: ....

Score: .... / 30

Addition & Subtraction (0-20)

①   1 2
 +   9
     . .
	② 1 8
	 - 1 2
	     . .
		③ 1 6
		 + 5
		     . .

① 1 2 + 9 . .

② 1 8 - 1 2 . .

③ 1 6 + 5 . .

④ 1 3 - 4 . .

⑤ 1 1 + 1 1 . .

⑥ 1 2 - 3 . .

⑦ 1 1 + 9 . .

⑧ 1 0 - 1 . .

⑨ 1 4 + 5 . .

⑩ 1 4 - 1 2 . .

⑪ 1 2 + 8 . .

⑫ 1 9 - 2 . .

⑬ 1 7 + 4 . .

⑭ 1 5 - 7 . .

⑮ 1 3 + 6 . .

⑯ 1 3 - 6 . .

⑰ 1 1 + 8 . .

⑱ 1 8 - 1 8 . .

⑲ 2 0 + 9 . .

⑳ 1 1 - 3 . .

㉑ 1 7 + 1 . .

㉒ 1 4 - 7 . .

㉓ 1 3 + 3 . .

㉔ 1 0 - 2 . .

㉕ 1 6 + 3 . .

㉖ 1 8 - 2 . .

㉗ 1 5 + 1 0 . .

㉘ 1 2 - 1 0 . .

㉙ 1 2 + 7 . .

㉚ 1 9 - 7 . .

Date: ....

Time: ....

Name: ....

Score: .... / 30

Addition & Subtraction (0-20)

1	$\begin{array}{r} 1\ 8 \\ -\ \ 3 \\ \hline \end{array}$	. .
2	$\begin{array}{r} 1\ 4 \\ +\ \ 7 \\ \hline \end{array}$	. .
3	$\begin{array}{r} 1\ 9 \\ -\ \ 6 \\ \hline \end{array}$	. .
4	$\begin{array}{r} 1\ 2 \\ +\ \ 3 \\ \hline \end{array}$	. .
5	$\begin{array}{r} 1\ 5 \\ -\ 1\ 2 \\ \hline \end{array}$	. .

6	$\begin{array}{r} 1\ 5 \\ +\ \ 4 \\ \hline \end{array}$	. .
7	$\begin{array}{r} 1\ 6 \\ -\ \ 1 \\ \hline \end{array}$	. .
8	$\begin{array}{r} 1\ 1 \\ +\ \ 2 \\ \hline \end{array}$	. .
9	$\begin{array}{r} 1\ 7 \\ -\ \ 5 \\ \hline \end{array}$	. .
10	$\begin{array}{r} 1\ 0 \\ +\ \ 5 \\ \hline \end{array}$	. .

11	$\begin{array}{r} 9 \\ +\ \ 6 \\ \hline \end{array}$	. .
12	$\begin{array}{r} 4 \\ -\ \ 2 \\ \hline \end{array}$	.
13	$\begin{array}{r} 1\ 6 \\ +\ \ 2 \\ \hline \end{array}$	. .
14	$\begin{array}{r} 5 \\ -\ \ 3 \\ \hline \end{array}$	.
15	$\begin{array}{r} 1\ 2 \\ +\ \ 5 \\ \hline \end{array}$	. .

16	$\begin{array}{r} 8 \\ -\ \ 2 \\ \hline \end{array}$	.
17	$\begin{array}{r} 1\ 0 \\ +\ \ 3 \\ \hline \end{array}$	. .
18	$\begin{array}{r} 6 \\ -\ \ 4 \\ \hline \end{array}$	.
19	$\begin{array}{r} 1\ 4 \\ +\ \ 1 \\ \hline \end{array}$	. .
20	$\begin{array}{r} 1\ 1 \\ -\ \ 5 \\ \hline \end{array}$	. .

21	$\begin{array}{r} 1\ 3 \\ +\ \ 4 \\ \hline \end{array}$	. .
22	$\begin{array}{r} 9 \\ -\ \ 2 \\ \hline \end{array}$	.
23	$\begin{array}{r} 1\ 5 \\ +\ \ 2 \\ \hline \end{array}$	. .
24	$\begin{array}{r} 7 \\ -\ \ 1 \\ \hline \end{array}$	.
25	$\begin{array}{r} 1\ 7 \\ +\ \ 2 \\ \hline \end{array}$	. .

26	$\begin{array}{r} 1\ 4 \\ -\ \ 1 \\ \hline \end{array}$	. .
27	$\begin{array}{r} 1\ 2 \\ +\ \ 6 \\ \hline \end{array}$	. .
28	$\begin{array}{r} 6 \\ -\ \ 2 \\ \hline \end{array}$	.
29	$\begin{array}{r} 1\ 0 \\ +\ \ 4 \\ \hline \end{array}$	.
30	$\begin{array}{r} 8 \\ -\ \ 5 \\ \hline \end{array}$	.

Date: .... Time: ....

Name: .... Score: .... / 30

Addition & Subtraction (0-20)

1	2	3	4	5
1 4 + 5 ..	5 - 1 .	1 1 + 1 ..	1 3 - 2 ..	1 4 + 3 ..

6	7	8	9	10
1 9 - 3 ..	1 5 + 4 ..	1 7 - 2 ..	1 7 + 3 ..	1 2 - 1 1 ..

11	12	13	14	15
1 8 + 2 ..	1 2 - 2 ..	1 3 + 5 ..	1 4 - 1 ..	1 1 + 3 ..

16	17	18	19	20
1 3 - 5 ..	1 9 + 9 ..	1 5 - 2 ..	1 4 + 1 2 ..	1 6 - 2 ..

21	22	23	24	25
1 2 + 4 ..	1 8 - 4 ..	1 0 + 1 ..	1 6 - 3 ..	1 4 + 6 ..

26	27	28	29	30
1 1 - 2 ..	2 0 + 7 ..	1 9 - 8 ..	1 9 + 1 ..	1 6 - 4 ..

Date: ....

Time: ....

Name: ....

Score: .... / 30

Addition & Subtraction (0-20)

1 $\begin{array}{r} 4 \\ +\ 7 \\ \hline . \end{array}$	**2** $\begin{array}{r} 9 \\ -\ 6 \\ \hline . \end{array}$	**3** $\begin{array}{r} 1\ 2 \\ +\ \ 3 \\ \hline . \end{array}$	**4** $\begin{array}{r} 5 \\ -\ 2 \\ \hline . \end{array}$	**5** $\begin{array}{r} 1\ 5 \\ +\ \ 4 \\ \hline .\ \ . \end{array}$
6 $\begin{array}{r} 6 \\ -\ 1 \\ \hline . \end{array}$	**7** $\begin{array}{r} 1\ 1 \\ +\ \ 2 \\ \hline . \end{array}$	**8** $\begin{array}{r} 7 \\ -\ 4 \\ \hline . \end{array}$	**9** $\begin{array}{r} 1\ 0 \\ +\ \ 5 \\ \hline . \end{array}$	**10** $\begin{array}{r} 3 \\ -\ 1 \\ \hline . \end{array}$
11 $\begin{array}{r} 1\ 3 \\ +\ \ 6 \\ \hline . \end{array}$	**12** $\begin{array}{r} 2 \\ -\ 1 \\ \hline . \end{array}$	**13** $\begin{array}{r} 1\ 4 \\ +\ \ 3 \\ \hline .\ \ . \end{array}$	**14** $\begin{array}{r} 7 \\ -\ 3 \\ \hline . \end{array}$	**15** $\begin{array}{r} 9 \\ +\ 6 \\ \hline . \end{array}$
16 $\begin{array}{r} 4 \\ -\ 2 \\ \hline . \end{array}$	**17** $\begin{array}{r} 1\ 6 \\ +\ \ 2 \\ \hline .\ \ . \end{array}$	**18** $\begin{array}{r} 5 \\ -\ 3 \\ \hline . \end{array}$	**19** $\begin{array}{r} 1\ 2 \\ +\ \ 5 \\ \hline . \end{array}$	**20** $\begin{array}{r} 8 \\ -\ 2 \\ \hline . \end{array}$
21 $\begin{array}{r} 1\ 0 \\ +\ \ 3 \\ \hline . \end{array}$	**22** $\begin{array}{r} 1\ 6 \\ -\ \ 4 \\ \hline .\ \ . \end{array}$	**23** $\begin{array}{r} 1\ 4 \\ +\ \ 1 \\ \hline .\ \ . \end{array}$	**24** $\begin{array}{r} 1\ 1 \\ -\ \ 5 \\ \hline . \end{array}$	**25** $\begin{array}{r} 1\ 3 \\ +\ \ 4 \\ \hline . \end{array}$
26 $\begin{array}{r} 9 \\ -\ 2 \\ \hline . \end{array}$	**27** $\begin{array}{r} 1\ 5 \\ +\ \ 2 \\ \hline .\ \ . \end{array}$	**28** $\begin{array}{r} 1\ 6 \\ -\ \ 4 \\ \hline .\ \ . \end{array}$	**29** $\begin{array}{r} 1\ 5 \\ +\ \ 3 \\ \hline .\ \ . \end{array}$	**30** $\begin{array}{r} 4 \\ -\ 1 \\ \hline . \end{array}$

Date: ....

Time: ....

Name: ....

Score: .... / 30

Addition & Subtraction (0-20)

1)
```
   1 2
+    6
-------
   . .
```

2)
```
     9
-    2
-------
     .
```

3)
```
   1 0
+    4
-------
   . .
```

4)
```
     8
-    5
-------
     .
```

5)
```
   1 4
+    5
-------
   . .
```

6)
```
     5
-    1
-------
     .
```

7)
```
   1 1
+    1
-------
   . .
```

8)
```
     3
-    2
-------
     .
```

9)
```
   1 3
+    3
-------
   . .
```

10)
```
     9
-    3
-------
     .
```

11)
```
   1 5
+    1
-------
   . .
```

12)
```
   1 7
-    2
-------
   . .
```

13)
```
   1 6
+    4
-------
   . .
```

14)
```
   1 2
-    4
-------
   . .
```

15)
```
   2 0
+    5
-------
   . .
```

16)
```
   1 0
-    1
-------
     .
```

17)
```
   1 7
+    2
-------
   . .
```

18)
```
   1 4
-    1
-------
   . .
```

19)
```
   1 1
+    3
-------
   . .
```

20)
```
   1 3
-    5
-------
   . .
```

21)
```
   1 9
+    1
-------
   . .
```

22)
```
   1 5
-    3
-------
   . .
```

23)
```
   1 7
+    2
-------
   . .
```

24)
```
   1 6
-    2
-------
   . .
```

25)
```
   1 2
+    2
-------
   . .
```

26)
```
   1 8
-    4
-------
   . .
```

27)
```
   1 0
+    1
-------
   . .
```

28)
```
   1 6
-    3
-------
   . .
```

29)
```
   1 4
+    2
-------
   . .
```

30)
```
   1 1
-    2
-------
   . .
```

Date: ....

Time: ....

Name: ....

Score: .... / 30

Addition & Subtraction (0-20)

1) $\begin{array}{r} 1\ 3 \\ +\quad 5 \\ \hline \end{array}$	2) $\begin{array}{r} 1\ 9 \\ -\quad 1 \\ \hline \end{array}$	3) $\begin{array}{r} 1\ 5 \\ +\quad 3 \\ \hline \end{array}$

1) 1 3 + 5

2) 1 9 − 1

3) 1 5 + 3

4) 1 7 − 4

5) 1 6 + 6

6) 1 2 − 5

7) 8 + 6

8) 1 8 − 9

9) 5 + 9

10) 1 6 − 3

11) 1 0 + 3

12) 1 9 − 7

13) 6 + 8

14) 1 5 − 7

15) 9 + 2

16) 1 7 − 4

17) 7 + 5

18) 2 0 − 6

19) 1 2 + 4

20) 1 4 − 2

21) 8 + 3

22) 1 9 − 6

23) 1 5 + 2

24) 1 6 − 8

25) 1 0 + 2

26) 1 7 − 6

27) 6 + 5

28) 1 5 − 3

29) 1 9 + 6

30) 1 7 − 8

Date:

Name:

Time:

Score: _____ / 30

Addition & Subtraction (0-20)

(1) 7 + 4	(2) 2 0 - 9	(3) 1 2 + 3	(4) 1 4 - 3	(5) 8 + 5
(6) 1 9 - 5	(7) 5 + 7	(8) 1 6 - 4	(9) 1 0 + 4	(10) 1 8 - 6
(11) 6 + 9	(12) 1 5 - 8	(13) 9 + 3	(14) 1 7 - 6	(15) 7 + 3
(16) 2 0 - 7	(17) 1 2 + 6	(18) 1 4 - 1	(19) 8 + 4	(20) 1 9 - 9
(21) 5 + 8	(22) 1 6 - 2	(23) 1 0 + 5	(24) 1 8 - 5	(25) 6 + 7
(26) 1 5 - 5	(27) 9 + 5	(28) 1 7 - 7	(29) 7 + 6	(30) 2 0 - 8

Date: ....

Time: ....

Name: ....

Score: .... / 30

Addition & Subtraction (0-20)

1	2	3	4	5
1 2 + 2	1 4 - 4	8 + 2	1 9 - 4	5 + 3

6	7	8	9	10
1 6 - 7	1 0 + 1	1 8 - 8	6 + 4	1 5 - 3

11	12	13	14	15
9 + 1	1 7 - 3	7 + 2	2 0 - 6	1 2 + 5

16	17	18	19	20
1 4 - 6	8 + 1	1 9 - 3	5 + 6	1 6 - 6

21	22	23	24	25
1 0 + 6	1 8 - 2	6 + 3	1 5 - 8	9 + 4

26	27	28	29	30
1 7 - 5	7 + 1	2 0 - 5	1 2 + 1	1 4 - 5

Date: ....

Time: ....

Name: ....

Score: .... / 30

Addition & Subtraction (0-20)

(1) $\begin{array}{r} 7 \\ + \ 4 \\ \hline \end{array}$	(2) $\begin{array}{r} 1\ 2 \\ - \ 6 \\ \hline \end{array}$	(3) $\begin{array}{r} 5 \\ + \ 2 \\ \hline \end{array}$	(4) $\begin{array}{r} 1\ 8 \\ - \ 9 \\ \hline \end{array}$	(5) $\begin{array}{r} 1\ 0 \\ + \ 3 \\ \hline \end{array}$
(6) $\begin{array}{r} 1\ 4 \\ - \ 8 \\ \hline \end{array}$	(7) $\begin{array}{r} 1\ 3 \\ + \ 6 \\ \hline \end{array}$	(8) $\begin{array}{r} 2\ 0 \\ - 1\ 0 \\ \hline \end{array}$	(9) $\begin{array}{r} 8 \\ + \ 5 \\ \hline \end{array}$	(10) $\begin{array}{r} 1\ 6 \\ - \ 7 \\ \hline \end{array}$
(11) $\begin{array}{r} 6 \\ + 1\ 1 \\ \hline \end{array}$	(12) $\begin{array}{r} 1\ 3 \\ - \ 4 \\ \hline \end{array}$	(13) $\begin{array}{r} 4 \\ + 1\ 5 \\ \hline \end{array}$	(14) $\begin{array}{r} 1\ 7 \\ - \ 9 \\ \hline \end{array}$	(15) $\begin{array}{r} 9 \\ + \ 2 \\ \hline \end{array}$
(16) $\begin{array}{r} 1\ 5 \\ - \ 6 \\ \hline \end{array}$	(17) $\begin{array}{r} 2 \\ + \ 8 \\ \hline \end{array}$	(18) $\begin{array}{r} 1\ 9 \\ - 1\ 1 \\ \hline \end{array}$	(19) $\begin{array}{r} 1\ 1 \\ + \ 4 \\ \hline \end{array}$	(20) $\begin{array}{r} 2\ 0 \\ - \ 8 \\ \hline \end{array}$
(21) $\begin{array}{r} 7 \\ + 1\ 2 \\ \hline \end{array}$	(22) $\begin{array}{r} 1\ 2 \\ - \ 5 \\ \hline \end{array}$	(23) $\begin{array}{r} 5 \\ + 1\ 4 \\ \hline \end{array}$	(24) $\begin{array}{r} 1\ 8 \\ - \ 7 \\ \hline \end{array}$	(25) $\begin{array}{r} 1\ 0 \\ + \ 1 \\ \hline \end{array}$
(26) $\begin{array}{r} 1\ 4 \\ - \ 6 \\ \hline \end{array}$	(27) $\begin{array}{r} 1\ 3 \\ + \ 3 \\ \hline \end{array}$	(28) $\begin{array}{r} 1\ 9 \\ - \ 7 \\ \hline \end{array}$	(29) $\begin{array}{r} 1\ 8 \\ + 1\ 1 \\ \hline \end{array}$	(30) $\begin{array}{r} 1\ 6 \\ - \ 4 \\ \hline \end{array}$

Date: ....

Time: ....

Name: ....

Score: .... / 30

Addition & Subtraction (0-20)

(1) 1 6 + 1 2 . .	(2) 1 3 - 3 . .	(3) 1 4 + 6 . .	(4) 1 7 - 8 . .	(5) 1 9 + 1 . .
(6) 1 5 - 5 . .	(7) 1 2 + 7 . .	(8) 1 9 - 1 0 . .	(9) 1 1 + 3 . .	(10) 2 0 - 9 . .
(11) 7 + 1 3 . .	(12) 1 2 - 1 . .	(13) 1 5 + 4 . .	(14) 1 8 - 6 . .	(15) 1 0 + 2 . .
(16) 1 4 - 7 . .	(17) 3 + 1 4 . .	(18) 2 0 - 1 1 . .	(19) 1 8 + 4 . .	(20) 1 6 - 6 . .
(21) 6 + 3 .	(22) 1 3 - 2 . .	(23) 4 + 1 7 . .	(24) 1 7 - 7 . .	(25) 9 + 3 . .
(26) 1 5 - 4 . .	(27) 1 2 + 9 . .	(28) 1 9 - 8 . .	(29) 1 1 + 2 . .	(30) 2 0 - 7 . .

Date: ....

Time: ....

Name: ....

Score: .... / 30

Addition & Subtraction (0-20)

1 $\begin{array}{r} 7 \\ +\ 1\ 1 \\ \hline \end{array}$	**2** $\begin{array}{r} 1\ 2 \\ -\ \ \ 4 \\ \hline \end{array}$	**3** $\begin{array}{r} 5 \\ +\ \ \ 6 \\ \hline \end{array}$	**4** $\begin{array}{r} 1\ 8 \\ -\ \ \ 5 \\ \hline \end{array}$	**5** $\begin{array}{r} 1\ 0 \\ +\ \ \ 4 \\ \hline \end{array}$
6 $\begin{array}{r} 1\ 4 \\ -\ \ \ 9 \\ \hline \end{array}$	**7** $\begin{array}{r} 4 \\ +\ 1\ 5 \\ \hline \end{array}$	**8** $\begin{array}{r} 2\ 0 \\ -\ \ \ 6 \\ \hline \end{array}$	**9** $\begin{array}{r} 8 \\ +\ \ \ 2 \\ \hline \end{array}$	**10** $\begin{array}{r} 1\ 6 \\ -\ \ \ 4 \\ \hline \end{array}$
11 $\begin{array}{r} 6 \\ +\ \ \ 4 \\ \hline \end{array}$	**12** $\begin{array}{r} 1\ 3 \\ -\ \ \ 5 \\ \hline \end{array}$	**13** $\begin{array}{r} 4 \\ +\ \ \ 8 \\ \hline \end{array}$	**14** $\begin{array}{r} 1\ 7 \\ -\ \ \ 5 \\ \hline \end{array}$	**15** $\begin{array}{r} 9 \\ +\ \ \ 4 \\ \hline \end{array}$
16 $\begin{array}{r} 1\ 5 \\ -\ \ \ 3 \\ \hline \end{array}$	**17** $\begin{array}{r} 2 \\ +\ 1\ 6 \\ \hline \end{array}$	**18** $\begin{array}{r} 1\ 9 \\ -\ \ \ 9 \\ \hline \end{array}$	**19** $\begin{array}{r} 1\ 1 \\ +\ \ \ 1 \\ \hline \end{array}$	**20** $\begin{array}{r} 2\ 0 \\ -\ \ \ 5 \\ \hline \end{array}$
21 $\begin{array}{r} 7 \\ +\ \ \ 5 \\ \hline \end{array}$	**22** $\begin{array}{r} 1\ 2 \\ -\ \ \ 2 \\ \hline \end{array}$	**23** $\begin{array}{r} 5 \\ +\ \ \ 7 \\ \hline \end{array}$	**24** $\begin{array}{r} 1\ 8 \\ -\ \ \ 4 \\ \hline \end{array}$	**25** $\begin{array}{r} 1\ 0 \\ +\ \ \ 5 \\ \hline \end{array}$
26 $\begin{array}{r} 1\ 4 \\ -\ \ \ 3 \\ \hline \end{array}$	**27** $\begin{array}{r} 1\ 3 \\ +\ \ \ 2 \\ \hline \end{array}$	**28** $\begin{array}{r} 2\ 0 \\ -\ \ \ 4 \\ \hline \end{array}$	**29** $\begin{array}{r} 1\ 8 \\ +\ \ \ 1 \\ \hline \end{array}$	**30** $\begin{array}{r} 1\ 6 \\ -\ \ \ 3 \\ \hline \end{array}$

Date: ....

Time: ....

Name: ....

Score: .... / 30

Addition & Subtraction (0-20)

1	2	3	4	5
1 1 + 3 . .	1 4 - 4 . .	9 + 6 . .	1 9 - 4 . .	8 + 4 . .
6	**7**	**8**	**9**	**10**
1 5 - 1 . .	9 + 5 . .	1 8 - 2 . .	8 + 6 . .	2 0 - 3 . .
11	**12**	**13**	**14**	**15**
1 3 + 4 . .	1 7 - 1 . .	1 0 + 7 . .	1 6 - 4 . .	9 + 9 . .
16	**17**	**18**	**19**	**20**
1 4 - 1 . .	6 + 5 . .	1 8 - 4 . .	8 + 5 . .	1 5 - 2 . .
21	**22**	**23**	**24**	**25**
4 + 4 .	1 3 - 1 . .	6 + 4 . .	2 0 - 2 0 . .	1 2 + 2 . .
26	**27**	**28**	**29**	**30**
2 0 - 1 . .	1 2 + 1 . .	1 8 - 8 . .	1 0 + 9 . .	1 5 - 5 . .

Date: ....

Time: ....

Name: ....

Score: .... / 30

Addition & Subtraction (0-20)

1 1 8 + 4	**2** 2 0 − 7	**3** 1 2 + 9

1
  1 8
+   4

2
  2 0
−   7

3
  1 2
+   9

4
  1 9
−   9

5
  1 4
+   6

6
  2 0
−   8

7
  1 5
+ 1 0

8
  1 7
−   5

9
  1 9
+   3

10
  2 0
− 1 1

11
  1 7
+   7

12
  2 0
−   6

13
  1 2
+ 1 4

14
  2 0
−   4

15
  1 6
+   5

16
  1 9
−   2

17
  1 8
+   2

18
  2 0
−   9

19
  2 0
+   9

20
  1 9
−   8

21
  1 6
+   4

22
  1 8
−   3

23
  1 9
+ 1 0

24
  2 0
−   7

25
  1 8
+   7

26
  1 7
−   4

27
  1 9
+   6

28
  2 0
−   5

29
  1 3
+ 1 7

30
  2 0
−   1

Date: ....

Time: ....

Name: ....

Score: ..../ 30

Addition & Subtraction (0-20)

(1) $\begin{array}{r} 19 \\ +11 \\ \hline \end{array}$	(2) $\begin{array}{r} 20 \\ -\ 2 \\ \hline \end{array}$	(3) $\begin{array}{r} 19 \\ +\ 8 \\ \hline \end{array}$	(4) $\begin{array}{r} 20 \\ -\ 7 \\ \hline \end{array}$	(5) $\begin{array}{r} 20 \\ +\ 1 \\ \hline \end{array}$
(6) $\begin{array}{r} 20 \\ -\ 9 \\ \hline \end{array}$	(7) $\begin{array}{r} 15 \\ +\ 5 \\ \hline \end{array}$	(8) $\begin{array}{r} 20 \\ -\ 3 \\ \hline \end{array}$	(9) $\begin{array}{r} 18 \\ +13 \\ \hline \end{array}$	(10) $\begin{array}{r} 20 \\ -10 \\ \hline \end{array}$
(11) $\begin{array}{r} 17 \\ +\ 3 \\ \hline \end{array}$	(12) $\begin{array}{r} 20 \\ -12 \\ \hline \end{array}$	(13) $\begin{array}{r} 19 \\ +14 \\ \hline \end{array}$	(14) $\begin{array}{r} 19 \\ -\ 4 \\ \hline \end{array}$	(15) $\begin{array}{r} 13 \\ +\ 8 \\ \hline \end{array}$
(16) $\begin{array}{r} 20 \\ -\ 5 \\ \hline \end{array}$	(17) $\begin{array}{r} 20 \\ +12 \\ \hline \end{array}$	(18) $\begin{array}{r} 20 \\ -\ 7 \\ \hline \end{array}$	(19) $\begin{array}{r} 4 \\ +\ 4 \\ \hline \end{array}$	(20) $\begin{array}{r} 20 \\ -\ 6 \\ \hline \end{array}$
(21) $\begin{array}{r} 20 \\ +\ 8 \\ \hline \end{array}$	(22) $\begin{array}{r} 19 \\ -\ 3 \\ \hline \end{array}$	(23) $\begin{array}{r} 16 \\ +\ 6 \\ \hline \end{array}$	(24) $\begin{array}{r} 10 \\ -\ 2 \\ \hline \end{array}$	(25) $\begin{array}{r} 20 \\ +11 \\ \hline \end{array}$
(26) $\begin{array}{r} 20 \\ -15 \\ \hline \end{array}$	(27) $\begin{array}{r} 17 \\ +\ 4 \\ \hline \end{array}$	(28) $\begin{array}{r} 19 \\ -11 \\ \hline \end{array}$	(29) $\begin{array}{r} 15 \\ +\ 9 \\ \hline \end{array}$	(30) $\begin{array}{r} 20 \\ -11 \\ \hline \end{array}$

Date: ....

Time: ....

Name: ....

Score: .... / 30

Addition & Subtraction (0-20)

① $\begin{array}{r} 1\ 3 \\ +\ 2\ 0 \\ \hline \end{array}$	② $\begin{array}{r} 2\ 0 \\ -\quad 8 \\ \hline \end{array}$	③ $\begin{array}{r} 1\ 6 \\ +\quad 7 \\ \hline \end{array}$

① $\begin{array}{r} 1\ 3 \\ +\ 2\ 0 \\ \hline \end{array}$ ② $\begin{array}{r} 2\ 0 \\ -\ 8 \\ \hline \end{array}$ ③ $\begin{array}{r} 1\ 6 \\ +\ 7 \\ \hline \end{array}$ ④ $\begin{array}{r} 2\ 0 \\ -\ 2 \\ \hline \end{array}$ ⑤ $\begin{array}{r} 2\ 0 \\ +\ 1\ 0 \\ \hline \end{array}$

⑥ $\begin{array}{r} 2\ 0 \\ -\ 6 \\ \hline \end{array}$ ⑦ $\begin{array}{r} 1\ 9 \\ +\ 4 \\ \hline \end{array}$ ⑧ $\begin{array}{r} 2\ 0 \\ -\ 5 \\ \hline \end{array}$ ⑨ $\begin{array}{r} 1\ 5 \\ +\ 6 \\ \hline \end{array}$ ⑩ $\begin{array}{r} 2\ 0 \\ -\ 4 \\ \hline \end{array}$

⑪ $\begin{array}{r} 1\ 2 \\ +\ 1\ 8 \\ \hline \end{array}$ ⑫ $\begin{array}{r} 2\ 0 \\ -\ 6 \\ \hline \end{array}$ ⑬ $\begin{array}{r} 1\ 7 \\ +\ 8 \\ \hline \end{array}$ ⑭ $\begin{array}{r} 1\ 0 \\ -\ 5 \\ \hline \end{array}$ ⑮ $\begin{array}{r} 1\ 4 \\ +\ 5 \\ \hline \end{array}$

⑯ $\begin{array}{r} 2\ 0 \\ -\ 4 \\ \hline \end{array}$ ⑰ $\begin{array}{r} 1\ 2 \\ +\ 9 \\ \hline \end{array}$ ⑱ $\begin{array}{r} 2\ 0 \\ -\ 2 \\ \hline \end{array}$ ⑲ $\begin{array}{r} 2\ 0 \\ +\ 3 \\ \hline \end{array}$ ⑳ $\begin{array}{r} 1\ 8 \\ -\ 8 \\ \hline \end{array}$

㉑ $\begin{array}{r} 2\ 0 \\ +\ 1\ 4 \\ \hline \end{array}$ ㉒ $\begin{array}{r} 2\ 0 \\ -\ 1\ 4 \\ \hline \end{array}$ ㉓ $\begin{array}{r} 1\ 2 \\ +\ 7 \\ \hline \end{array}$ ㉔ $\begin{array}{r} 2\ 0 \\ -\ 1 \\ \hline \end{array}$ ㉕ $\begin{array}{r} 1\ 5 \\ +\ 1\ 1 \\ \hline \end{array}$

㉖ $\begin{array}{r} 2\ 0 \\ -\ 3 \\ \hline \end{array}$ ㉗ $\begin{array}{r} 1\ 3 \\ +\ 4 \\ \hline \end{array}$ ㉘ $\begin{array}{r} 2\ 0 \\ -\ 2 \\ \hline \end{array}$ ㉙ $\begin{array}{r} 1\ 2 \\ +\ 6 \\ \hline \end{array}$ ㉚ $\begin{array}{r} 2\ 0 \\ -\ 1 \\ \hline \end{array}$

Date: ....

Time: ....

Name: ....

Score: .... / 30

Addition & Subtraction (0-20)

1. $\begin{array}{r} 17 \\ +\ \ 4 \\ \hline \end{array}$	2. $\begin{array}{r} 20 \\ -\ \ 6 \\ \hline \end{array}$	3. $\begin{array}{r} 13 \\ +\ \ 7 \\ \hline \end{array}$	4. $\begin{array}{r} 18 \\ -\ \ 9 \\ \hline \end{array}$	5. $\begin{array}{r} 15 \\ +\ \ 7 \\ \hline \end{array}$
6. $\begin{array}{r} 19 \\ -\ \ 7 \\ \hline \end{array}$	7. $\begin{array}{r} 16 \\ +11 \\ \hline \end{array}$	8. $\begin{array}{r} 20 \\ -\ \ 5 \\ \hline \end{array}$	9. $\begin{array}{r} 18 \\ +\ \ 3 \\ \hline \end{array}$	10. $\begin{array}{r} 20 \\ -10 \\ \hline \end{array}$
11. $\begin{array}{r} 15 \\ +\ \ 8 \\ \hline \end{array}$	12. $\begin{array}{r} 20 \\ -\ \ 7 \\ \hline \end{array}$	13. $\begin{array}{r} 17 \\ +12 \\ \hline \end{array}$	14. $\begin{array}{r} 20 \\ -\ \ 3 \\ \hline \end{array}$	15. $\begin{array}{r} 14 \\ +\ \ 6 \\ \hline \end{array}$
16. $\begin{array}{r} 20 \\ -\ \ 2 \\ \hline \end{array}$	17. $\begin{array}{r} 19 \\ +\ \ 2 \\ \hline \end{array}$	18. $\begin{array}{r} 20 \\ -\ \ 8 \\ \hline \end{array}$	19. $\begin{array}{r} 20 \\ +10 \\ \hline \end{array}$	20. $\begin{array}{r} 20 \\ -\ \ 9 \\ \hline \end{array}$
21. $\begin{array}{r} 4 \\ +15 \\ \hline \end{array}$	22. $\begin{array}{r} 20 \\ -\ \ 4 \\ \hline \end{array}$	23. $\begin{array}{r} 19 \\ +\ \ 7 \\ \hline \end{array}$	24. $\begin{array}{r} 20 \\ -\ \ 6 \\ \hline \end{array}$	25. $\begin{array}{r} 16 \\ +13 \\ \hline \end{array}$
26. $\begin{array}{r} 20 \\ -\ \ 4 \\ \hline \end{array}$	27. $\begin{array}{r} 18 \\ +\ \ 5 \\ \hline \end{array}$	28. $\begin{array}{r} 20 \\ -\ \ 7 \\ \hline \end{array}$	29. $\begin{array}{r} 19 \\ +\ \ 9 \\ \hline \end{array}$	30. $\begin{array}{r} 20 \\ -\ \ 1 \\ \hline \end{array}$

Date: ....

Time: ....

Name: ....

Score: .... / 30

Addition & Subtraction (0-20)

(1) $\begin{array}{r} 1\ 1 \\ +\ 1\ 8 \\ \hline \end{array}$	(2) $\begin{array}{r} 2\ 0 \\ -\ \ \ 3 \\ \hline \end{array}$	(3) $\begin{array}{r} 1\ 7 \\ +\ \ \ 8 \\ \hline \end{array}$	(4) $\begin{array}{r} 2\ 0 \\ -\ \ \ 5 \\ \hline \end{array}$	(5) $\begin{array}{r} 2\ 0 \\ +\ \ \ 1 \\ \hline \end{array}$
(6) $\begin{array}{r} 2\ 0 \\ -\ \ \ 8 \\ \hline \end{array}$	(7) $\begin{array}{r} 1\ 6 \\ +\ 1\ 6 \\ \hline \end{array}$	(8) $\begin{array}{r} 2\ 0 \\ -\ \ \ 3 \\ \hline \end{array}$	(9) $\begin{array}{r} 1\ 9 \\ +\ 1\ 0 \\ \hline \end{array}$	(10) $\begin{array}{r} 2\ 0 \\ -\ \ \ 9 \\ \hline \end{array}$
(11) $\begin{array}{r} 1\ 7 \\ +\ \ \ 3 \\ \hline \end{array}$	(12) $\begin{array}{r} 2\ 0 \\ -\ 1\ 5 \\ \hline \end{array}$	(13) $\begin{array}{r} 1\ 4 \\ +\ 1\ 9 \\ \hline \end{array}$	(14) $\begin{array}{r} 2\ 0 \\ -\ \ \ 4 \\ \hline \end{array}$	(15) $\begin{array}{r} 1\ 3 \\ +\ \ \ 7 \\ \hline \end{array}$
(16) $\begin{array}{r} 1\ 6 \\ -\ \ \ 5 \\ \hline \end{array}$	(17) $\begin{array}{r} 2\ 0 \\ +\ 1\ 2 \\ \hline \end{array}$	(18) $\begin{array}{r} 2\ 0 \\ -\ \ \ 7 \\ \hline \end{array}$	(19) $\begin{array}{r} 1\ 4 \\ +\ \ \ 4 \\ \hline \end{array}$	(20) $\begin{array}{r} 2\ 0 \\ -\ \ \ 6 \\ \hline \end{array}$
(21) $\begin{array}{r} 8 \\ +\ 2\ 0 \\ \hline \end{array}$	(22) $\begin{array}{r} 1\ 9 \\ -\ \ \ 3 \\ \hline \end{array}$	(23) $\begin{array}{r} 1\ 6 \\ +\ \ \ 6 \\ \hline \end{array}$	(24) $\begin{array}{r} 2\ 0 \\ -\ \ \ 2 \\ \hline \end{array}$	(25) $\begin{array}{r} 1\ 1 \\ +\ 2\ 0 \\ \hline \end{array}$
(26) $\begin{array}{r} 1\ 9 \\ -\ \ \ 5 \\ \hline \end{array}$	(27) $\begin{array}{r} 2\ 0 \\ +\ \ \ 5 \\ \hline \end{array}$	(28) $\begin{array}{r} 2\ 0 \\ -\ 1\ 4 \\ \hline \end{array}$	(29) $\begin{array}{r} 1\ 5 \\ +\ \ \ 9 \\ \hline \end{array}$	(30) $\begin{array}{r} 1\ 8 \\ -\ \ \ 3 \\ \hline \end{array}$

Day 47

Addition & Subtraction (0-20)

1 13 + 2 0 . .	2 2 0 - 8 . .	3 1 6 + 7 . .	4 2 0 - 2 . .	5 2 0 + 1 0 . .
6 2 0 - 6 . .	7 1 9 + 4 . .	8 2 0 - 5 . .	9 1 5 + 6 . .	10 1 8 - 8 . .
11 1 9 + 1 2 . .	12 2 0 - 6 . .	13 1 7 + 8 . .	14 1 7 - 7 . .	15 1 4 + 5 . .
16 1 3 - 4 . .	17 1 2 + 9 . .	18 1 8 - 9 . .	19 2 0 + 3 . .	20 1 7 - 6 . .
21 1 4 + 2 0 . .	22 1 9 - 9 . .	23 1 2 + 6 . .	24 2 0 - 1 . .	25 1 3 + 2 4 . .
26 1 9 - 6 . .	27 1 8 + 6 . .	28 1 6 - 5 . .	29 1 7 + 1 3 . .	30 2 0 - 7 . .

Addition & Subtraction (0-20)

① 1 9 - 8 . .	② 1 1 + 1 8 . .	③ 9 + 1 0 . .	④ 1 3 - 4 . .	⑤ 1 8 - 7 . .
⑥ 7 + 9 . .	⑦ 1 5 - 3 . .	⑧ 1 9 + 1 2 . .	⑨ 1 7 - 5 . .	⑩ 1 0 + 1 4 . .
⑪ 1 6 - 6 . .	⑫ 6 + 7 . .	⑬ 1 9 - 9 . .	⑭ 1 4 + 1 6 . .	⑮ 1 0 + 5 . .
⑯ 1 8 - 6 . .	⑰ 1 3 + 8 . .	⑱ 1 7 - 4 . .	⑲ 1 1 + 1 5 . .	⑳ 1 9 - 7 . .
㉑ 9 + 1 2 . .	㉒ 1 6 - 5 . .	㉓ 1 7 + 1 3 . .	㉔ 1 1 + 7 . .	㉕ 1 9 - 6 . .
㉖ 1 0 + 1 3 . .	㉗ 1 8 - 4 . .	㉘ 1 4 + 1 5 . .	㉙ 1 0 + 8 . .	㉚ 2 0 - 9 . .

Date: ....

Time: ....

Name: ....

Score: .... / 30

Addition & Subtraction (0-20)

1)
```
    1 2
  + 1 4
  -------
    . .
```

2)
```
    1 7
  -   3
  -------
    . .
```

3)
```
      9
  +   6
  -------
    . .
```

4)
```
    1 6
  -   7
  -------
    . .
```

5)
```
    1 3
  + 1 5
  -------
    . .
```

6)
```
      9
  +   8
  -------
    . .
```

7)
```
    1 9
  -   4
  -------
    . .
```

8)
```
    1 1
  + 1 3
  -------
    . .
```

9)
```
    1 8
  -   3
  -------
    . .
```

10)
```
    1 1
  + 1 0
  -------
    . .
```

11)
```
    1 5
  -   6
  -------
    . .
```

12)
```
      7
  +   8
  -------
    . .
```

13)
```
    1 9
  -   3
  -------
    . .
```

14)
```
    1 4
  + 1 2
  -------
    . .
```

15)
```
    1 6
  -   4
  -------
    . .
```

16)
```
    1 3
  + 1 4
  -------
    . .
```

17)
```
      6
  +   6
  -------
    . .
```

18)
```
    1 9
  -   2
  -------
    . .
```

19)
```
    1 0
  +   9
  -------
    . .
```

20)
```
    1 8
  -   5
  -------
    . .
```

21)
```
    1 2
  + 1 1
  -------
    . .
```

22)
```
    1 5
  -   7
  -------
    . .
```

23)
```
      7
  +   8
  -------
    . .
```

24)
```
    1 9
  -   1
  -------
    . .
```

25)
```
    1 2
  + 1 1
  -------
    . .
```

26)
```
    1 7
  -   6
  -------
    . .
```

27)
```
      9
  +   6
  -------
    . .
```

28)
```
    1 4
  +   9
  -------
    . .
```

29)
```
    1 9
  -   5
  -------
    . .
```

30)
```
    1 3
  - 1 2
  -------
    . .
```

Date: ....

Time: ....

Name: ....

Score: .... / 30

Addition & Subtraction (0-20)

①	②	③	④	⑤
1 6 - 3 . .	7 + 5 . .	1 9 - 1 1 . .	1 0 + 8 . .	1 8 - 7 . .

⑥	⑦	⑧	⑨	⑩
1 1 + 1 0 . .	1 5 - 5 . .	8 + 6 . .	1 9 - 2 . .	1 2 + 1 0 . .

⑪	⑫	⑬	⑭	⑮
1 7 - 4 . .	9 + 5 . .	1 4 + 8 . .	1 9 - 1 . .	1 3 + 1 1 . .

⑯	⑰	⑱	⑲	⑳
1 6 - 6 . .	7 + 4 . .	1 8 - 3 . .	1 1 + 9 . .	1 5 - 4 . .

㉑	㉒	㉓	㉔	㉕
8 + 5 . .	1 9 - 1 . .	1 0 + 7 . .	1 7 - 5 . .	1 2 + 9 . .

㉖	㉗	㉘	㉙	㉚
1 4 + 7 . .	1 9 - 4 . .	1 3 + 1 0 . .	1 6 - 2 . .	7 + 3 . .

Date: ....

Time: ....

Name: ....

Score: .... / 30

Addition & Subtraction (2 Digit)

① 1 5 + 2 5 . .	② 4 5 - 1 0 . .	③ 2 0 + 3 1 . .	④ 5 1 - 2 0 . .	⑤ 3 5 + 1 0 . .
⑥ 5 5 - 1 0 . .	⑦ 2 5 + 2 2 . .	⑧ 6 0 - 3 0 . .	⑨ 3 2 + 1 1 . .	⑩ 4 3 - 1 0 . .
⑪ 4 6 + 1 3 . .	⑫ 6 5 - 2 0 . .	⑬ 1 0 + 4 1 . .	⑭ 5 0 - 3 0 . .	⑮ 2 1 + 2 1 . .
⑯ 7 2 - 3 0 . .	⑰ 2 5 + 2 5 . .	⑱ 8 8 - 2 0 . .	⑲ 3 4 + 1 5 . .	⑳ 5 5 - 2 0 . .
㉑ 1 5 + 4 3 . .	㉒ 7 2 - 4 1 . .	㉓ 2 5 + 1 0 . .	㉔ 5 5 - 2 5 . .	㉕ 2 3 + 3 0 . .
㉖ 6 0 - 4 0 . .	㉗ 3 0 + 1 2 . .	㉘ 5 4 - 2 4 . .	㉙ 1 0 + 3 3 . .	㉚ 3 6 - 1 3 . .

Addition & Subtraction (2 Digit)

①
```
    3 3
 +  1 5
 _____
  . .
```

②
```
    7 7
 -  3 0
 _____
  . .
```

③
```
    2 8
 +  2 1
 _____
  . .
```

④
```
    6 0
 -  2 0
 _____
  . .
```

⑤
```
    2 5
 +  2 3
 _____
  . .
```

⑥
```
    8 0
 -  2 0
 _____
  . .
```

⑦
```
    3 5
 +  1 1
 _____
  . .
```

⑧
```
    6 3
 -  4 0
 _____
  . .
```

⑨
```
    1 3
 +  1 5
 _____
  . .
```

⑩
```
    9 9
 -  1 0
 _____
  . .
```

⑪
```
    7 2
 +  1 0
 _____
  . .
```

⑫
```
    3 5
 -  1 5
 _____
  . .
```

⑬
```
    3 0
 +  1 8
 _____
  . .
```

⑭
```
    5 5
 -  2 4
 _____
  . .
```

⑮
```
    3 1
 +  4 4
 _____
  . .
```

⑯
```
    7 0
 -  4 0
 _____
  . .
```

⑰
```
    2 8
 +  1 1
 _____
  . .
```

⑱
```
    5 3
 -  1 2
 _____
  . .
```

⑲
```
    1 4
 +  1 5
 _____
  . .
```

⑳
```
    3 6
 -  3 0
 _____
  . .
```

㉑
```
    6 5
 +  3 3
 _____
  . .
```

㉒
```
    9 5
 -  7 4
 _____
  . .
```

㉓
```
    4 3
 +  2 1
 _____
  . .
```

㉔
```
    7 8
 -  1 5
 _____
  . .
```

㉕
```
    1 0
 +  7 9
 _____
  . .
```

㉖
```
    4 2
 -  3 1
 _____
  . .
```

㉗
```
    8 0
 +  1 1
 _____
  . .
```

㉘
```
    7 6
 -  4 5
 _____
  . .
```

㉙
```
    1 8
 +  6 1
 _____
  . .
```

㉚
```
    7 8
 -  5 5
 _____
  . .
```

Date: ....

Name: ....

Time: ....

Score: .... / 30

Addition & Subtraction (2 Digit)

1)
```
    4 2
  + 2 3
  _____
    . .
```

2)
```
    8 9
  - 3 2
  _____
    . .
```

3)
```
    1 1
  + 3 1
  _____
    . .
```

4)
```
    6 8
  - 3 7
  _____
    . .
```

5)
```
    5 1
  + 1 7
  _____
    . .
```

6)
```
    9 7
  - 4 3
  _____
    . .
```

7)
```
    1 9
  + 3 5
  _____
    . .
```

8)
```
    7 6
  - 3 4
  _____
    . .
```

9)
```
    2 7
  + 4 1
  _____
    . .
```

10)
```
    8 6
  - 2 3
  _____
    . .
```

11)
```
    1 3
  + 1 9
  _____
    . .
```

12)
```
    6 9
  - 2 6
  _____
    . .
```

13)
```
    3 5
  + 4 2
  _____
    . .
```

14)
```
    9 5
  - 4 4
  _____
    . .
```

15)
```
    2 3
  + 3 6
  _____
    . .
```

16)
```
    5 8
  - 2 5
  _____
    . .
```

17)
```
    1 7
  + 1 6
  _____
    . .
```

18)
```
    6 7
  - 2 3
  _____
    . .
```

19)
```
    1 4
  + 2 8
  _____
    . .
```

20)
```
    7 8
  - 4 7
  _____
    . .
```

21)
```
    2 9
  + 3 8
  _____
    . .
```

22)
```
    8 5
  - 4 1
  _____
    . .
```

23)
```
    3 3
  + 1 9
  _____
    . .
```

24)
```
    7 6
  - 2 2
  _____
    . .
```

25)
```
    2 6
  + 4 5
  _____
    . .
```

26)
```
    9 7
  - 5 2
  _____
    . .
```

27)
```
    1 2
  + 3 6
  _____
    . .
```

28)
```
    7 5
  - 3 5
  _____
    . .
```

29)
```
    2 1
  + 3 1
  _____
    . .
```

30)
```
    6 5
  - 3 5
  _____
    . .
```

Date: .... Time: ....

Name: .... Score: .... / 30

Addition & Subtraction (2 Digit)

① 1 8 + 4 5	② 8 7 - 2 4	③ 3 6 + 1 5	④ 5 6 - 3 5	⑤ 1 6 + 1 8
⑥ 4 7 - 2 6	⑦ 2 1 + 2 4	⑧ 6 8 - 3 2	⑨ 2 3 + 1 6	⑩ 7 9 - 4 8
⑪ 1 5 + 1 5	⑫ 6 8 - 1 5	⑬ 2 8 + 4 1	⑭ 9 7 - 3 1	⑮ 3 7 + 1 3
⑯ 7 5 - 2 4	⑰ 1 4 + 2 8	⑱ 5 8 - 2 4	⑲ 2 6 + 4 8	⑳ 8 9 - 5 6
㉑ 1 3 + 1 2	㉒ 6 4 - 3 3	㉓ 3 2 + 2 8	㉔ 6 8 - 1 4	㉕ 1 9 + 1 4
㉖ 4 9 - 1 7	㉗ 1 7 + 4 8	㉘ 7 9 - 2 6	㉙ 3 1 + 6 8	㉚ 9 9 - 3 9

Date: ....

Time: ....

Name: ....

Score: .... / 30

Addition & Subtraction (2 Digit)

(1) 22 + 1 3 . .	(2) 88 - 2 3 . .	(3) 2 6 + 4 5 . .	(4) 8 9 - 5 5 . .	(5) 1 9 + 2 4 . .
(6) 5 7 - 3 4 . .	(7) 1 1 + 2 8 . .	(8) 4 8 - 1 6 . .	(9) 3 3 + 1 7 . .	(10) 7 8 - 3 2 . .
(11) 1 3 + 5 5 . .	(12) 7 7 - 3 2 . .	(13) 2 5 + 4 5 . .	(14) 8 6 - 2 2 . .	(15) 1 6 + 4 3 . .
(16) 9 9 - 5 9 . .	(17) 1 2 + 1 9 . .	(18) 5 9 - 2 7 . .	(19) 1 8 + 1 7 . .	(20) 4 9 - 2 8 . .
(21) 2 3 + 2 4 . .	(22) 7 8 - 3 6 . .	(23) 1 7 + 3 1 . .	(24) 6 9 - 1 7 . .	(25) 2 9 + 1 6 . .
(26) 8 6 - 4 2 . .	(27) 3 2 + 1 7 . .	(28) 5 9 - 3 3 . .	(29) 2 4 + 2 9 . .	(30) 7 6 - 2 4 . .

Date: ....

Time: ....

Name: ....

Score: .... / 30

Addition & Subtraction (2 Digit)

①
```
  1 4
+ 3 2
------
. .
```

②
```
  6 2
- 5 5
------
. .
```

③
```
  3 1
+ 2 6
------
. .
```

④
```
  7 8
- 3 7
------
. .
```

⑤
```
  1 5
+ 2 3
------
. .
```

⑥
```
  4 9
- 1 6
------
. .
```

⑦
```
  1 8
+ 4 6
------
. .
```

⑧
```
  9 5
- 4 9
------
. .
```

⑨
```
  1 1
+ 2 8
------
. .
```

⑩
```
  5 9
- 3 8
------
. .
```

⑪
```
  2 7
+ 3 5
------
. .
```

⑫
```
  8 5
- 3 9
------
. .
```

⑬
```
  1 2
+ 2 8
------
. .
```

⑭
```
  7 7
- 4 5
------
. .
```

⑮
```
  2 8
+ 1 9
------
. .
```

⑯
```
  6 4
- 2 8
------
. .
```

⑰
```
  1 9
+ 2 1
------
. .
```

⑱
```
  7 8
- 3 9
------
. .
```

⑲
```
  2 5
+ 4 6
------
. .
```

⑳
```
  9 3
- 5 5
------
. .
```

㉑
```
  1 8
+ 1 4
------
. .
```

㉒
```
  4 9
- 1 8
------
. .
```

㉓
```
  1 5
+ 1 9
------
. .
```

㉔
```
  6 5
- 2 6
------
. .
```

㉕
```
  2 3
+ 2 4
------
. .
```

㉖
```
  7 7
- 4 3
------
. .
```

㉗
```
  2 7
+ 3 9
------
. .
```

㉘
```
  8 1
- 3 8
------
. .
```

㉙
```
  2 2
+ 1 4
------
. .
```

㉚
```
  6 8
- 3 2
------
. .
```

Date: ....

Time: ....

Name: ....

Score: .... / 30

Addition & Subtraction (2 Digit)

1	3 4 + 4 6 . .	
2	9 5 - 4 5 . .	
3	1 9 + 1 7 . .	
4	5 4 - 2 8 . .	
5	1 8 + 3 1 . .	

1. 3 4
   + 4 6
    . .

2. 9 5
   - 4 5
    . .

3. 1 9
   + 1 7
    . .

4. 5 4
   - 2 8
    . .

5. 1 8
   + 3 1
    . .

6. 6 9
   - 3 5
    . .

7. 2 5
   + 3 8
    . .

8. 9 2
   - 5 5
    . .

9. 1 4
   + 1 9
    . .

10. 7 5
    - 2 7
     . .

11. 2 9
    + 4 5
     . .

12. 9 6
    - 3 8
     . .

13. 2 2
    + 1 4
     . .

14. 5 8
    - 2 6
     . .

15. 1 8
    + 3 8
     . .

16. 7 7
    - 3 5
     . .

17. 2 3
    + 2 4
     . .

18. 6 7
    - 4 9
     . .

19. 1 6
    + 1 9
     . .

20. 5 4
    - 2 9
     . .

21. 2 7
    + 1 5
     . .

22. 6 6
    - 2 8
     . .

23. 1 7
    + 6 6
     . .

24. 5 8
    - 2 8
     . .

25. 4 3
    + 3 5
     . .

26. 7 2
    - 3 8
     . .

27. 1 6
    + 4 9
     . .

28. 9 7
    - 4 2
     . .

29. 1 8
    + 1 7
     . .

30. 6 3
    - 2 9
     . .

Day 58

Date:

Name:

Time:

Score: ___ / 30

Addition & Subtraction (2 Digit)

① 29 + 16

② 78 − 39

③ 13 + 25

④ 64 − 26

⑤ 27 + 21

⑥ 71 − 25

⑦ 21 + 16

⑧ 48 − 29

⑨ 18 + 35

⑩ 86 − 37

⑪ 23 + 28

⑫ 59 − 35

⑬ 16 + 42

⑭ 84 − 45

⑮ 12 + 16

⑯ 47 − 23

⑰ 22 + 37

⑱ 79 − 38

⑲ 14 + 31

⑳ 66 − 35

㉑ 24 + 17

㉒ 85 − 46

㉓ 16 + 31

㉔ 88 − 37

㉕ 23 + 38

㉖ 73 − 35

㉗ 18 + 45

㉘ 86 − 38

㉙ 13 + 19

㉚ 47 − 28

Addition & Subtraction (2 Digit)

1 2 5 + 2 7	**2** 6 8 - 3 8	**3** 1 8 + 1 5	**4** 4 3 - 2 4	**5** 1 4 + 2 5
6 5 2 - 2 6	**7** 1 9 + 3 2	**8** 6 5 - 1 9	**9** 2 9 + 2 8	**10** 7 1 - 4 5
11 4 3 + 1 9	**12** 6 8 - 2 8	**13** 2 2 + 3 8	**14** 7 5 - 3 6	**15** 1 4 + 1 9
16 5 8 - 2 5	**17** 1 5 + 4 7	**18** 8 4 - 2 4	**19** 2 7 + 1 8	**20** 5 9 - 2 5
21 2 1 + 2 8	**22** 7 6 - 3 9	**23** 1 9 + 3 7	**24** 9 1 - 1 8	**25** 2 6 + 1 4
26 4 2 - 2 7	**27** 1 3 + 3 6	**28** 5 6 - 2 4	**29** 2 1 + 2 4	**30** 7 1 - 2 9

Date: ....

Time: ....

Name: ....

Score: .... / 30

Addition & Subtraction (2 Digit)

① 2 7 + 3 7 . .	② 8 5 - 3 9 . .	③ 1 2 + 2 4 . .	④ 6 4 - 3 5 . .	⑤ 1 9 + 1 7 . .
⑥ 4 3 - 2 6 . .	⑦ 1 7 + 2 8 . .	⑧ 6 9 - 3 8 . .	⑨ 2 4 + 2 7 . .	⑩ 8 7 - 4 2 . .
⑪ 1 6 + 1 5 . .	⑫ 5 3 - 2 7 . .	⑬ 2 5 + 1 8 . .	⑭ 7 9 - 4 8 . .	⑮ 1 3 + 2 8 . .
⑯ 7 2 - 2 8 . .	⑰ 2 8 + 4 1 . .	⑱ 8 5 - 3 8 . .	⑲ 1 8 + 1 7 . .	⑳ 6 7 - 2 8 . .
㉑ 3 1 + 1 8 . .	㉒ 6 9 - 3 5 . .	㉓ 1 8 + 2 8 . .	㉔ 6 2 - 3 5 . .	㉕ 2 4 + 1 8 . .
㉖ 9 1 - 3 5 . .	㉗ 1 6 + 3 1 . .	㉘ 5 3 - 2 9 . .	㉙ 1 3 + 1 6 . .	㉚ 4 8 - 2 6 . .

Date:

Time:

Name:

Score: ____ / 30

Addition & Subtraction (2 Digit)

1)
```
  1 2
+ 1 9
-----
  . .
```

2)
```
  4 3
- 2 4
-----
  . .
```

3)
```
  1 4
+ 3 6
-----
  . .
```

4)
```
  6 9
- 3 5
-----
  . .
```

5)
```
  2 7
+ 2 4
-----
  . .
```

6)
```
  8 7
- 3 9
-----
  . .
```

7)
```
  1 3
+ 1 5
-----
  . .
```

8)
```
  5 4
- 2 9
-----
  . .
```

9)
```
  2 3
+ 3 8
-----
  . .
```

10)
```
  7 5
- 3 8
-----
  . .
```

11)
```
  1 7
+ 3 5
-----
  . .
```

12)
```
  6 9
- 3 2
-----
  . .
```

13)
```
  2 6
+ 2 9
-----
  . .
```

14)
```
  7 6
- 3 9
-----
  . .
```

15)
```
  1 3
+ 2 1
-----
  . .
```

16)
```
  4 8
- 2 3
-----
  . .
```

17)
```
  1 6
+ 2 8
-----
  . .
```

18)
```
  6 2
- 2 7
-----
  . .
```

19)
```
  1 9
+ 1 8
-----
  . .
```

20)
```
  4 6
- 2 8
-----
  . .
```

21)
```
  2 7
+ 1 9
-----
  . .
```

22)
```
  8 5
- 3 5
-----
  . .
```

23)
```
  1 4
+ 2 5
-----
  . .
```

24)
```
  5 9
- 2 8
-----
  . .
```

25)
```
  2 3
+ 3 8
-----
  . .
```

26)
```
  7 1
- 3 8
-----
  . .
```

27)
```
  1 6
+ 3 9
-----
  . .
```

28)
```
  7 8
- 3 2
-----
  . .
```

29)
```
  1 9
+ 1 5
-----
  . .
```

30)
```
  4 3
- 2 3
-----
  . .
```

Date: ....

Time: ....

Name: ....

Score: .... / 30

Addition & Subtraction (2 Digit)

(1) 5 6 + 2 3	(2) 8 7 - 3 4	(3) 4 3 + 2 1	(4) 7 5 - 1 6	(5) 6 6 + 1 3
(6) 9 5 - 2 2	(7) 4 8 + 3 7	(8) 8 2 - 4 1	(9) 5 4 + 1 2	(10) 6 8 - 2 3
(11) 3 2 + 4 5	(12) 9 9 - 3 8	(13) 5 9 + 2 8	(14) 7 4 - 1 9	(15) 4 1 + 3 4
(16) 8 8 - 4 3	(17) 6 3 + 1 1	(18) 9 1 - 1 6	(19) 5 2 + 1 7	(20) 7 7 - 3 5
(21) 3 6 + 4 2	(22) 9 8 - 1 7	(23) 6 2 + 1 9	(24) 8 3 - 2 5	(25) 5 7 + 1 5
(26) 9 2 - 3 1	(27) 4 5 + 2 8	(28) 7 9 - 2 4	(29) 6 1 + 1 8	(30) 8 4 - 2 2

Date: ....

Time: ....

Name: ....

Score: .... / 30

Addition & Subtraction (2 Digit)

① $\begin{array}{r} 3\ 3 \\ +\ 2\ 7 \\ \hline \end{array}$ . .	② $\begin{array}{r} 9\ 5 \\ -\ 6\ 3 \\ \hline \end{array}$ . .	③ $\begin{array}{r} 4\ 6 \\ +\ 1\ 4 \\ \hline \end{array}$ . .

④ $\begin{array}{r} 7\ 6 \\ -\ 2\ 9 \\ \hline \end{array}$ . .

⑤ $\begin{array}{r} 3\ 8 \\ +\ 3\ 3 \\ \hline \end{array}$ . .

⑥ $\begin{array}{r} 8\ 9 \\ -\ 4\ 2 \\ \hline \end{array}$ . .

⑦ $\begin{array}{r} 6\ 4 \\ +\ 1\ 6 \\ \hline \end{array}$ . .

⑧ $\begin{array}{r} 9\ 3 \\ -\ 1\ 7 \\ \hline \end{array}$ . .

⑨ $\begin{array}{r} 5\ 8 \\ +\ 1\ 8 \\ \hline \end{array}$ . .

⑩ $\begin{array}{r} 7\ 2 \\ -\ 4\ 3 \\ \hline \end{array}$ . .

⑪ $\begin{array}{r} 5\ 1 \\ +\ 2\ 2 \\ \hline \end{array}$ . .

⑫ $\begin{array}{r} 8\ 5 \\ -\ 3\ 9 \\ \hline \end{array}$ . .

⑬ $\begin{array}{r} 4\ 7 \\ +\ 2\ 6 \\ \hline \end{array}$ . .

⑭ $\begin{array}{r} 7\ 1 \\ -\ 2\ 7 \\ \hline \end{array}$ . .

⑮ $\begin{array}{r} 3\ 9 \\ +\ 3\ 2 \\ \hline \end{array}$ . .

⑯ $\begin{array}{r} 9\ 7 \\ -\ 6\ 4 \\ \hline \end{array}$ . .

⑰ $\begin{array}{r} 5\ 3 \\ +\ 1\ 6 \\ \hline \end{array}$ . .

⑱ $\begin{array}{r} 8\ 8 \\ -\ 3\ 6 \\ \hline \end{array}$ . .

⑲ $\begin{array}{r} 6\ 6 \\ +\ 1\ 9 \\ \hline \end{array}$ . .

⑳ $\begin{array}{r} 9\ 4 \\ -\ 2\ 9 \\ \hline \end{array}$ . .

㉑ $\begin{array}{r} 3\ 4 \\ +\ 4\ 7 \\ \hline \end{array}$ . .

㉒ $\begin{array}{r} 7\ 6 \\ -\ 4\ 2 \\ \hline \end{array}$ . .

㉓ $\begin{array}{r} 4\ 9 \\ +\ 2\ 3 \\ \hline \end{array}$ . .

㉔ $\begin{array}{r} 6\ 9 \\ -\ 2\ 8 \\ \hline \end{array}$ . .

㉕ $\begin{array}{r} 4\ 2 \\ +\ 3\ 6 \\ \hline \end{array}$ . .

㉖ $\begin{array}{r} 9\ 0 \\ -\ 4\ 4 \\ \hline \end{array}$ . .

㉗ $\begin{array}{r} 6\ 7 \\ +\ 1\ 8 \\ \hline \end{array}$ . .

㉘ $\begin{array}{r} 9\ 6 \\ -\ 2\ 2 \\ \hline \end{array}$ . .

㉙ $\begin{array}{r} 5\ 5 \\ +\ 1\ 2 \\ \hline \end{array}$ . .

㉚ $\begin{array}{r} 8\ 0 \\ -\ 3\ 8 \\ \hline \end{array}$ . .

Date: ....

Name: ....

Time: ....

Score: .... / 30

Addition & Subtraction (2 Digit)

①
```
   3 5
 + 1 4
 ------
  . .
```

②
```
   7 4
 - 4 8
 ------
  . .
```

③
```
   5 6
 + 1 7
 ------
  . .
```

④
```
   8 1
 - 1 2
 ------
  . .
```

⑤
```
   4 3
 + 2 3
 ------
  . .
```

⑥
```
   9 8
 - 6 2
 ------
  . .
```

⑦
```
   5 9
 + 1 6
 ------
  . .
```

⑧
```
   8 6
 - 2 4
 ------
  . .
```

⑨
```
   6 3
 + 1 4
 ------
  . .
```

⑩
```
   9 2
 - 3 7
 ------
  . .
```

⑪
```
   3 1
 + 4 5
 ------
  . .
```

⑫
```
   7 3
 - 2 6
 ------
  . .
```

⑬
```
   5 7
 + 1 3
 ------
  . .
```

⑭
```
   7 8
 - 2 4
 ------
  . .
```

⑮
```
   4 0
 + 3 6
 ------
  . .
```

⑯
```
   8 7
 - 4 9
 ------
  . .
```

⑰
```
   6 4
 + 1 9
 ------
  . .
```

⑱
```
   9 3
 - 5 1
 ------
  . .
```

⑲
```
   4 8
 + 1 5
 ------
  . .
```

⑳
```
   8 2
 - 2 9
 ------
  . .
```

㉑
```
   3 7
 + 1 6
 ------
  . .
```

㉒
```
   7 5
 - 4 1
 ------
  . .
```

㉓
```
   5 4
 + 1 3
 ------
  . .
```

㉔
```
   7 9
 - 2 8
 ------
  . .
```

㉕
```
   4 4
 + 2 7
 ------
  . .
```

㉖
```
   9 6
 - 6 5
 ------
  . .
```

㉗
```
   6 1
 + 1 5
 ------
  . .
```

㉘
```
   8 9
 - 4 7
 ------
  . .
```

㉙
```
   6 5
 + 1 1
 ------
  . .
```

㉚
```
   9 1
 - 1 9
 ------
  . .
```

Date: ....

Time: ....

Name: ....

Score: .... / 30

Addition & Subtraction (2 Digit)

1	2	3	4	5
2 1 + 2 7	6 8 - 2 6	1 7 + 1 9	3 9 - 2 6	1 8 + 2 9

6	7	8	9	10
5 6 - 2 6	2 3 + 1 7	7 1 - 3 8	1 9 + 2 8	7 9 - 4 2

11	12	13	14	15
1 2 + 2 4	5 9 - 2 8	1 6 + 2 9	6 8 - 3 8	2 3 + 1 4

16	17	18	19	20
4 8 - 2 3	1 9 + 2 8	7 3 - 2 9	2 7 + 1 7	6 6 - 2 9

21	22	23	24	25
1 7 + 1 9	4 5 - 2 8	1 5 + 3 7	5 7 - 2 5	2 2 + 2 8

26	27	28	29	30
7 5 - 3 7	1 5 + 2 6	6 4 - 3 5	1 9 + 2 7	7 7 - 4 5

Date: ....

Time: ....

Name: ....

Score: .... / 30

Addition & Subtraction (2 Digit)

1	2	3	4	5
3 2 + 3 6	8 3 - 4 1	5 8 + 1 7	7 0 - 2 6	4 5 + 2 9

6	7	8	9	10
7 7 - 3 5	6 2 + 1 8	9 4 - 4 5	5 7 + 1 4	8 8 - 3 2

11	12	13	14	15
4 5 + 2 3	7 8 - 3 6	1 4 + 3 9	9 1 - 1 8	5 6 + 2 7

16	17	18	19	20
8 5 - 4 4	3 2 + 5 8	9 7 - 2 2	7 6 + 1 5	8 3 - 5 9

21	22	23	24	25
2 0 + 5 8	7 7 - 1 2	6 1 + 3 6	9 2 - 2 5	4 3 + 5 4

26	27	28	29	30
8 8 - 1 1	5 9 + 3 1	6 6 - 2 2	3 9 + 2 6	8 4 - 4 2

Date: ....

Time: ....

Name: ....

Score: .... / 30

Addition & Subtraction (2 Digit)

1)
```
  1 7
+ 5 8
------
  . .
```

2)
```
  9 5
- 2 7
------
  . .
```

3)
```
  5 1
+ 4 4
------
  . .
```

4)
```
  8 9
- 2 8
------
  . .
```

5)
```
  7 0
+ 2 6
------
  . .
```

6)
```
  7 4
- 1 8
------
  . .
```

7)
```
  3 3
+ 4 3
------
  . .
```

8)
```
  9 3
- 2 1
------
  . .
```

9)
```
  2 2
+ 6 7
------
  . .
```

10)
```
  8 1
- 2 9
------
  . .
```

11)
```
  6 2
+ 3 7
------
  . .
```

12)
```
  8 6
- 4 8
------
  . .
```

13)
```
  4 8
+ 4 2
------
  . .
```

14)
```
  7 4
- 3 1
------
  . .
```

15)
```
  5 3
+ 4 5
------
  . .
```

16)
```
  8 9
- 3 3
------
  . .
```

17)
```
  3 7
+ 2 6
------
  . .
```

18)
```
  9 2
- 1 6
------
  . .
```

19)
```
  6 7
+ 1 5
------
  . .
```

20)
```
  8 1
- 5 5
------
  . .
```

21)
```
  5 5
+ 2 3
------
  . .
```

22)
```
  8 3
- 2 8
------
  . .
```

23)
```
  3 8
+ 4 5
------
  . .
```

24)
```
  9 7
- 1 2
------
  . .
```

25)
```
  6 4
+ 2 1
------
  . .
```

26)
```
  7 5
- 4 2
------
  . .
```

27)
```
  2 6
+ 4 3
------
  . .
```

28)
```
  9 4
- 2 5
------
  . .
```

29)
```
  1 6
+ 6 7
------
  . .
```

30)
```
  8 2
- 3 3
------
  . .
```

Addition & Subtraction (2 Digit)

1 3 3 + 6 4	**2** 7 7 − 1 3	**3** 4 2 + 3 6	**4** 6 8 − 2 7	**5** 1 9 + 3 8
6 8 5 − 4 5	**7** 5 4 + 1 4	**8** 9 1 − 3 9	**9** 2 9 + 5 5	**10** 7 8 − 2 6
11 4 6 + 2 3	**12** 9 5 − 5 8	**13** 3 9 + 2 4	**14** 7 2 − 2 8	**15** 5 4 + 4 2
16 8 6 − 1 9	**17** 2 5 + 5 9	**18** 8 3 − 4 7	**19** 4 9 + 2 6	**20** 7 6 − 1 4
21 3 4 + 4 5	**22** 9 2 − 3 5	**23** 4 8 + 1 8	**24** 6 2 − 2 7	**25** 2 5 + 5 5
26 8 7 − 1 3	**27** 3 5 + 4 4	**28** 9 6 − 3 2	**29** 1 4 + 6 4	**30** 8 4 − 4 9

Date:

Time:

Name:

Score: .... / 30

Addition & Subtraction (2 Digit)

1	2	3	4	5
5 4 + 2 5 . .	7 3 - 2 1 . .	2 8 + 4 9 . .	9 7 - 6 5 . .	1 7 + 3 8 . .

6	7	8	9	10
6 2 - 2 8 . .	4 5 + 3 6 . .	8 5 - 4 1 . .	3 2 + 4 9 . .	9 3 - 2 2 . .

11	12	13	14	15
5 7 + 3 4 . .	7 9 - 4 3 . .	3 9 + 3 5 . .	6 5 - 2 7 . .	1 8 + 7 6 . .

16	17	18	19	20
8 1 - 3 4 . .	5 6 + 2 3 . .	9 4 - 2 8 . .	3 1 + 6 8 . .	8 8 - 4 6 . .

21	22	23	24	25
3 9 + 1 7 . .	8 2 - 4 2 . .	7 6 + 1 2 . .	6 8 - 2 2 . .	2 7 + 3 8 . .

26	27	28	29	30
9 5 - 1 3 . .	2 4 + 6 3 . .	4 1 - 1 5 . .	8 4 + 1 1 . .	7 2 - 4 8 . .

Date: ....

Time: ....

Name: ....

Score: .... / 30

Addition & Subtraction (2 Digit)

① 35 + 42	② 78 - 25	③ 57 + 22

① 35
+ 42
. .

② 78
- 25
. .

③ 57
+ 22
. .

④ 86
- 43
. .

⑤ 62
+ 18
. .

⑥ 91
- 27
. .

⑦ 40
+ 15
. .

⑧ 72
- 30
. .

⑨ 46
+ 23
. .

⑩ 88
- 16
. .

⑪ 24
+ 33
. .

⑫ 95
- 42
. .

⑬ 53
+ 16
. .

⑭ 82
- 19
. .

⑮ 68
+ 11
. .

⑯ 97
- 38
. .

⑰ 39
+ 17
. .

⑱ 73
- 24
. .

⑲ 51
+ 14
. .

⑳ 89
- 33
. .

㉑ 26
+ 38
. .

㉒ 94
- 16
. .

㉓ 60
+ 14
. .

㉔ 87
- 33
. .

㉕ 43
+ 18
. .

㉖ 77
- 34
. .

㉗ 31
+ 28
. .

㉘ 98
- 46
. .

㉙ 63
+ 27
. .

㉚ 92
- 15
. .

Addition & Subtraction (2 Digit)

(1) 47 + 2 6 . .	(2) 6 2 - 3 5 . .	(3) 1 8 + 7 9 . .	(4) 8 1 - 4 3 . .	(5) 1 5 + 3 9 . .
(6) 9 4 - 5 8 . .	(7) 2 3 + 4 6 . .	(8) 7 5 - 2 9 . .	(9) 3 4 + 6 2 . .	(10) 9 8 - 5 7 . .
(11) 2 8 + 5 9 . .	(12) 8 3 - 4 7 . .	(13) 1 2 + 6 7 . .	(14) 8 8 - 4 9 . .	(15) 3 6 + 4 2 . .
(16) 7 4 - 2 1 . .	(17) 5 7 + 1 8 . .	(18) 9 2 - 3 4 . .	(19) 2 1 + 3 3 . .	(20) 6 3 - 1 7 . .
(21) 4 5 + 3 2 . .	(22) 6 9 - 2 8 . .	(23) 2 7 + 6 4 . .	(24) 8 7 - 3 9 . .	(25) 3 8 + 4 9 . .
(26) 7 2 - 2 7 . .	(27) 3 1 + 5 6 . .	(28) 8 4 - 3 8 . .	(29) 1 9 + 5 8 . .	(30) 9 6 - 4 3 . .

Date: ....

Time: ....

Name: ....

Score: .... / 30

Addition & Subtraction (2 Digit)

1	2	3	4	5
5 3 + 2 6	7 7 - 1 8	2 4 + 3 1	6 6 - 3 5	3 9 + 2 3

6	7	8	9	10
9 1 - 5 4	1 7 + 4 2	8 3 - 3 8	6 1 + 2 8	9 5 - 5 4

11	12	13	14	15
4 7 + 2 1	7 2 - 3 1	2 9 + 4 9	8 6 - 3 8	3 7 + 2 4

16	17	18	19	20
7 9 - 4 7	2 6 + 6 8	9 4 - 3 2	3 1 + 4 7	5 7 - 2 2

21	22	23	24	25
5 6 + 2 3	7 7 - 1 2	1 9 + 3 4	6 8 - 2 4	5 2 + 2 1

26	27	28	29	30
8 6 - 3 3	3 6 + 4 9	9 1 - 2 3	1 3 + 4 8	7 4 - 1 6

Date:

Name:

Time:

Score: _____ / 30

Addition & Subtraction (2 Digit)

① 26 + 41 . .	② 83 − 27 . .	③ 57 + 18 . .	④ 94 − 44 . .	⑤ 38 + 29 . .
⑥ 62 − 11 . .	⑦ 27 + 37 . .	⑧ 59 − 15 . .	⑨ 81 + 18 . .	⑩ 98 − 49 . .
⑪ 43 + 24 . .	⑫ 77 − 42 . .	⑬ 25 + 45 . .	⑭ 85 − 13 . .	⑮ 39 + 19 . .
⑯ 67 − 22 . .	⑰ 14 + 58 . .	⑱ 82 − 48 . .	⑲ 49 + 18 . .	⑳ 76 − 23 . .
㉑ 41 + 31 . .	㉒ 69 − 27 . .	㉓ 18 + 36 . .	㉔ 97 − 55 . .	㉕ 24 + 15 . .
㉖ 83 − 26 . .	㉗ 54 + 14 . .	㉘ 86 − 35 . .	㉙ 47 + 23 . .	㉚ 72 − 29 . .

Date: ....

Time: ....

Name: ....

Score: .... / 30

Addition & Subtraction (2 Digit)

1)
```
   3 2
 + 4 7
 ------
   . .
```

2)
```
   9 3
 - 4 8
 ------
   . .
```

3)
```
   1 9
 + 2 8
 ------
   . .
```

4)
```
   7 7
 - 4 2
 ------
   . .
```

5)
```
   5 8
 + 1 5
 ------
   . .
```

6)
```
   8 9
 - 2 7
 ------
   . .
```

7)
```
   3 5
 + 4 9
 ------
   . .
```

8)
```
   9 6
 - 4 3
 ------
   . .
```

9)
```
   4 1
 + 2 4
 ------
   . .
```

10)
```
   8 5
 - 4 2
 ------
   . .
```

11)
```
   2 9
 + 3 9
 ------
   . .
```

12)
```
   7 4
 - 1 3
 ------
   . .
```

13)
```
   1 6
 + 2 8
 ------
   . .
```

14)
```
   6 9
 - 1 7
 ------
   . .
```

15)
```
   2 1
 + 3 8
 ------
   . .
```

16)
```
   8 8
 - 2 9
 ------
   . .
```

17)
```
   4 4
 + 2 3
 ------
   . .
```

18)
```
   9 7
 - 5 2
 ------
   . .
```

19)
```
   2 5
 + 1 6
 ------
   . .
```

20)
```
   6 6
 - 1 7
 ------
   . .
```

21)
```
   1 3
 + 4 9
 ------
   . .
```

22)
```
   9 2
 - 3 1
 ------
   . .
```

23)
```
   3 4
 + 4 7
 ------
   . .
```

24)
```
   7 8
 - 1 5
 ------
   . .
```

25)
```
   3 9
 + 2 7
 ------
   . .
```

26)
```
   5 9
 - 2 4
 ------
   . .
```

27)
```
   1 7
 + 3 7
 ------
   . .
```

28)
```
   8 3
 - 1 8
 ------
   . .
```

29)
```
   3 7
 + 2 4
 ------
   . .
```

30)
```
   9 6
 - 4 6
 ------
   . .
```

Date: ....   Time: ....

Name: ....   Score: .... / 30

Addition & Subtraction (2 Digit)

(1) 3 6 + 1 9	(2) 7 1 - 2 9	(3) 2 8 + 3 8	(4) 8 5 - 3 7	(5) 1 6 + 2 4
(6) 6 3 - 1 2	(7) 2 2 + 4 5	(8) 9 4 - 3 2	(9) 4 6 + 2 7	(10) 7 3 - 2 1
(11) 3 7 + 1 9	(12) 8 8 - 4 6	(13) 2 5 + 1 8	(14) 5 9 - 3 1	(15) 3 2 + 4 6
(16) 7 8 - 2 4	(17) 1 4 + 1 9	(18) 5 3 - 1 7	(19) 1 8 + 3 7	(20) 8 7 - 4 1
(21) 3 1 + 1 5	(22) 6 8 - 2 6	(23) 2 4 + 2 9	(24) 7 7 - 3 4	(25) 1 4 + 1 8
(26) 6 3 - 1 2	(27) 3 3 + 4 7	(28) 9 2 - 3 5	(29) 3 8 + 2 9	(30) 8 6 - 1 3

Date: ....

Time: ....

Name: ....

Score: .... / 30

Addition & Subtraction (3 Digit)

①	②	③	④	⑤
1 2 8 + 2 3 4 . . .	6 7 2 - 3 1 8 . . .	3 8 5 + 2 0 6 . . .	8 2 7 - 1 3 2 . . .	5 5 3 + 1 2 2 . . .
⑥	⑦	⑧	⑨	⑩
9 4 7 - 2 1 5 . . .	7 1 4 + 1 3 2 . . .	6 2 5 - 1 8 9 . . .	3 2 8 + 1 1 0 . . .	9 2 7 - 4 0 8 . . .
⑪	⑫	⑬	⑭	⑮
4 5 5 + 1 6 6 . . .	6 7 9 - 2 8 3 . . .	3 7 5 + 1 4 6 . . .	8 4 8 - 3 2 5 . . .	5 9 7 + 1 2 5 . . .
⑯	⑰	⑱	⑲	⑳
9 3 1 - 2 1 7 . . .	4 9 6 + 1 3 1 . . .	8 2 5 - 2 1 2 . . .	3 6 1 + 1 0 2 . . .	9 4 8 - 5 3 2 . . .
㉑	㉒	㉓	㉔	㉕
5 2 9 + 1 3 5 . . .	7 1 7 - 2 0 5 . . .	4 8 2 + 2 3 4 . . .	8 2 6 - 2 4 5 . . .	6 2 5 + 1 8 5 . . .
㉖	㉗	㉘	㉙	㉚
8 4 5 - 2 2 4 . . .	3 8 9 + 1 3 9 . . .	9 3 1 - 3 0 7 . . .	6 8 9 + 1 1 5 . . .	8 2 4 - 1 9 9 . . .

Date: ....

Time: ....

Name: ....

Score: .... / 30

Addition & Subtraction (3 Digit)

(1)	(2)	(3)	(4)	(5)
518 +187	712 -203	361 +142	938 -326	485 +211

(6)	(7)	(8)	(9)	(10)
816 -224	628 +119	932 -314	399 +186	841 -218

(11)	(12)	(13)	(14)	(15)
545 +126	948 -432	628 +142	715 -206	489 +204

(16)	(17)	(18)	(19)	(20)
913 -236	478 +155	927 -202	555 +121	819 -244

(21)	(22)	(23)	(24)	(25)
594 +113	826 -214	362 +133	931 -304	618 +157

(26)	(27)	(28)	(29)	(30)
894 -378	485 +123	716 -205	529 +147	948 -435

Date: ....

Time: ....

Name: ....

Score: .... / 30

Addition & Subtraction (3 Digit)

①	②	③	④	⑤
6 3 5	8 2 7	2 6 8	8 3 2	6 7 4
+ 2 1 0	- 2 1 6	+ 2 5 6	- 2 2 0	+ 1 0 4

⑥	⑦	⑧	⑨	⑩
9 1 5	5 8 6	7 2 9	3 8 7	7 1 4
- 2 0 7	+ 1 5 5	- 2 0 4	+ 3 1 5	- 1 1 5

⑪	⑫	⑬	⑭	⑮
5 3 8	6 4 8	6 5 5	8 2 6	3 9 8
+ 1 4 8	- 3 1 7	+ 3 1 0	- 2 3 8	+ 1 3 0

⑯	⑰	⑱	⑲	⑳
6 1 8	6 8 9	6 2 8	5 1 9	7 4 8
- 2 4 8	+ 1 3 2	- 3 1 6	+ 1 3 8	- 2 3 7

㉑	㉒	㉓	㉔	㉕
4 8 5	9 1 2	4 1 8	8 2 5	5 5 5
+ 2 4 1	- 3 0 4	+ 2 3 7	- 2 1 8	+ 1 1 5

㉖	㉗	㉘	㉙	㉚
8 1 9	6 7 9	6 3 8	4 8 9	7 2 6
- 3 1 7	+ 1 4 8	- 3 1 4	+ 1 2 0	- 2 1 6

Date: .... Time: ....

Name: .... Score: .... / 30

Addition & Subtraction (3 Digit)

1	2	3	4	5
1 8 4 + 2 4 7	5 7 2 - 3 2 9	3 9 8 + 2 1 4	8 1 7 - 3 2 5	4 1 9 + 2 3 8

6	7	8	9	10
9 2 5 - 3 0 8	7 2 6 + 1 6 2	8 5 6 - 3 1 2	4 7 8 + 1 8 3	6 2 9 - 2 3 8

11	12	13	14	15
3 4 1 + 2 0 9	8 6 8 - 2 9 3	6 3 4 + 1 8 3	6 2 6 - 2 0 7	5 6 4 + 2 1 8

16	17	18	19	20
9 1 6 - 3 4 7	6 9 3 + 2 7 8	9 2 7 - 2 7 9	6 2 7 + 2 1 7	8 9 5 - 3 7 8

21	22	23	24	25
5 3 2 + 1 6 3	8 5 8 - 2 8 3	4 7 9 + 3 3 1	9 3 8 - 3 7 9	6 4 7 + 1 5 2

26	27	28	29	30
9 7 8 - 2 7 6	5 9 2 + 2 1 6	9 2 6 - 3 8 2	5 4 1 + 1 7 3	8 7 4 - 2 3 8

Addition & Subtraction (3 Digit)

1)
```
  6 2 5
+ 1 9 3
```
· · ·

2)
```
  9 8 7
- 5 2 7
```
· · ·

3)
```
  6 5 3
+ 1 2 6
```
· · ·

4)
```
  8 1 2
- 2 3 2
```
· · ·

5)
```
  5 3 7
+ 2 1 4
```
· · ·

6)
```
  8 6 7
- 2 8 6
```
· · ·

7)
```
  6 5 4
+ 2 1 9
```
· · ·

8)
```
  7 2 5
- 3 1 8
```
· · ·

9)
```
  5 6 7
+ 2 1 8
```
· · ·

10)
```
  9 2 4
- 3 6 8
```
· · ·

11)
```
  6 9 8
+ 2 1 4
```
· · ·

12)
```
  7 4 5
- 1 8 5
```
· · ·

13)
```
  4 1 7
+ 2 6 4
```
· · ·

14)
```
  9 5 6
- 6 2 8
```
· · ·

15)
```
  7 1 2
+ 1 9 3
```
· · ·

16)
```
  8 2 9
- 2 3 7
```
· · ·

17)
```
  5 3 9
+ 3 1 7
```
· · ·

18)
```
  8 4 7
- 1 8 6
```
· · ·

19)
```
  5 2 4
+ 3 2 5
```
· · ·

20)
```
  9 2 8
- 4 7 2
```
· · ·

21)
```
  6 5 3
+ 2 7 7
```
· · ·

22)
```
  7 7 5
- 3 1 6
```
· · ·

23)
```
  5 9 1
+ 2 1 7
```
· · ·

24)
```
  9 1 4
- 2 8 4
```
· · ·

25)
```
  3 2 6
+ 2 1 9
```
· · ·

26)
```
  9 3 6
- 3 2 7
```
· · ·

27)
```
  2 1 4
+ 6 7 4
```
· · ·

28)
```
  8 2 8
- 2 7 6
```
· · ·

29)
```
  2 2 7
+ 5 1 4
```
· · ·

30)
```
  9 3 8
- 3 4 7
```
· · ·

Date: ....

Time: ....

Name: ....

Score: .... / 30

Addition & Subtraction (3 Digit)

(1) 514 + 236	(2) 827 − 328	(3) 549 + 192

(4) 924 − 327

(5) 654 + 215

(6) 856 − 329

(7) 612 + 107

(8) 928 − 287

(9) 643 + 117

(10) 845 − 175

(11) 569 + 193

(12) 925 − 317

(13) 632 + 219

(14) 917 − 328

(15) 561 + 218

(16) 937 − 289

(17) 647 + 219

(18) 859 − 387

(19) 612 + 218

(20) 937 − 327

(21) 538 + 193

(22) 867 − 278

(23) 628 + 318

(24) 728 − 327

(25) 656 + 317

(26) 875 − 426

(27) 543 + 111

(28) 976 − 238

(29) 615 + 192

(30) 995 − 238

Date: ....

Time: ....

Name: ....

Score: .... / 30

Addition & Subtraction (3 Digit)

① 444
+ 3 1 9
. . .

② 938
- 3 2 7
. . .

③ 272
+ 5 1 9
. . .

④ 879
- 5 8 9
. . .

⑤ 263
+ 5 7 6
. . .

⑥ 936
- 5 2 7
. . .

⑦ 519
+ 3 3 8
. . .

⑧ 649
- 3 4 7
. . .

⑨ 576
+ 2 2 3
. . .

⑩ 937
- 2 2 7
. . .

⑪ 541
+ 4 3 9
. . .

⑫ 917
- 4 0 8
. . .

⑬ 724
+ 1 8 5
. . .

⑭ 666
- 2 3 7
. . .

⑮ 130
+ 5 8 9
. . .

⑯ 872
- 2 3 6
. . .

⑰ 734
+ 1 5 7
. . .

⑱ 962
- 3 2 7
. . .

⑲ 712
+ 1 3 6
. . .

⑳ 927
- 2 9 6
. . .

㉑ 258
+ 1 7 4
. . .

㉒ 739
- 2 3 5
. . .

㉓ 576
+ 1 8 9
. . .

㉔ 942
- 3 1 7
. . .

㉕ 621
+ 2 3 8
. . .

㉖ 897
- 2 8 9
. . .

㉗ 647
+ 1 7 3
. . .

㉘ 926
- 3 0 7
. . .

㉙ 679
+ 2 3 1
. . .

㉚ 957
- 2 8 6
. . .

Date:

Time:

Name:

Score: .... / 30

Addition & Subtraction (3 Digit)

① 7 1 9 + 1 5 8	② 8 6 4 - 1 9 5	③ 2 5 2 + 1 7 4	④ 9 4 3 - 2 1 7	⑤ 5 9 6 + 1 8 9
⑥ 8 3 1 - 3 0 5	⑦ 1 3 4 + 7 5 7	⑧ 8 6 9 - 1 9 7	⑨ 5 4 3 + 1 7 8	⑩ 8 6 5 - 2 1 6
⑪ 6 3 8 + 1 9 8	⑫ 8 1 9 - 3 1 6	⑬ 5 9 7 + 1 3 6	⑭ 8 4 7 - 2 1 4	⑮ 1 6 8 + 6 9 3
⑯ 9 2 4 - 2 5 9	⑰ 6 5 7 + 1 2 8	⑱ 8 6 2 - 2 1 7	⑲ 1 4 3 + 7 6 7	⑳ 8 2 8 - 3 1 6
㉑ 6 8 4 + 2 7 2	㉒ 8 7 5 - 3 3 8	㉓ 6 5 9 + 2 2 2	㉔ 9 5 4 - 4 2 8	㉕ 7 2 3 + 1 1 6
㉖ 9 3 2 - 2 3 6	㉗ 5 9 6 + 1 2 9	㉘ 5 2 6 - 1 1 8	㉙ 1 9 2 + 6 3 8	㉚ 8 4 9 - 2 7 6

Date:

Time:

Name:

Score: .... / 30

Addition & Subtraction (3 Digit)

① 631 + 2 4 9 . . .	② 956 - 2 5 9 . . .	③ 627 + 1 7 8 . . .	④ 957 - 3 2 8 . . .	⑤ 689 + 1 5 7 . . .
⑥ 923 - 3 6 8 . . .	⑦ 238 + 6 1 7 . . .	⑧ 845 - 3 2 7 . . .	⑨ 652 + 1 3 6 . . .	⑩ 943 - 3 1 7 . . .
⑪ 189 + 5 9 6 . . .	⑫ 931 - 2 0 5 . . .	⑬ 734 + 1 5 8 . . .	⑭ 765 - 1 1 7 . . .	⑮ 643 + 1 0 0 . . .
⑯ 627 - 1 6 6 . . .	⑰ 136 + 7 1 4 . . .	⑱ 962 - 3 2 7 . . .	⑲ 734 + 1 5 7 . . .	⑳ 872 - 2 3 6 . . .
㉑ 198 + 6 3 8 . . .	㉒ 917 - 3 1 6 . . .	㉓ 136 + 5 9 7 . . .	㉔ 847 - 2 1 4 . . .	㉕ 144 + 6 9 3 . . .
㉖ 924 - 2 5 9 . . .	㉗ 128 + 6 5 7 . . .	㉘ 962 - 3 2 7 . . .	㉙ 167 + 7 4 3 . . .	㉚ 928 - 3 1 6 . . .

Date: ....

Time: ....

Name: ....

Score: .... / 30

Addition & Subtraction (3 Digit)

① 684 + 172	② 875 − 238	③ 659 + 189	④ 854 − 225	⑤ 723 + 178
⑥ 822 − 236	⑦ 596 + 119	⑧ 926 − 306	⑨ 692 + 238	⑩ 849 − 276
⑪ 731 + 159	⑫ 856 − 259	⑬ 178 + 627	⑭ 957 − 328	⑮ 157 + 689
⑯ 923 − 368	⑰ 238 + 617	⑱ 845 − 218	⑲ 136 + 652	⑳ 943 − 217
㉑ 596 + 289	㉒ 931 − 305	㉓ 164 + 726	㉔ 865 − 227	㉕ 643 + 186
㉖ 727 − 196	㉗ 136 + 712	㉘ 945 − 228	㉙ 285 + 418	㉚ 779 − 489

Date: ....

Time: ....

Name: ....

Score: .... / 30

Addition & Subtraction (3 Digit)

① 173 + 2 1 6 . . .	② 987 - 1 9 8 . . .	③ 342 + 1 3 5 . . .	④ 764 - 2 8 9 . . .	⑤ 421 + 2 7 3 . . .
⑥ 843 - 2 8 7 . . .	⑦ 548 + 1 4 6 . . .	⑧ 932 - 4 1 8 . . .	⑨ 672 + 1 8 2 . . .	⑩ 738 - 4 8 2 . . .
⑪ 563 + 2 1 4 . . .	⑫ 827 - 3 1 8 . . .	⑬ 714 + 1 3 7 . . .	⑭ 852 - 2 0 5 . . .	⑮ 624 + 1 4 8 . . .
⑯ 919 - 3 1 8 . . .	⑰ 567 + 2 1 8 . . .	⑱ 941 - 2 6 7 . . .	⑲ 694 + 1 7 3 . . .	⑳ 869 - 2 4 8 . . .
㉑ 698 + 2 3 1 . . .	㉒ 957 - 3 6 9 . . .	㉓ 748 + 1 7 2 . . .	㉔ 936 - 2 2 6 . . .	㉕ 619 + 2 4 5 . . .
㉖ 928 - 2 9 8 . . .	㉗ 675 + 1 6 6 . . .	㉘ 894 - 2 8 7 . . .	㉙ 541 + 1 9 8 . . .	㉚ 978 - 3 7 7 . . .

Date: ....

Time: ....

Name: ....

Score: .... / 30

Addition & Subtraction (3 Digit)

(1) 7 3 2 + 1 6 8	(2) 8 1 6 - 2 0 7	(3) 5 8 9 + 1 9 4	(4) 9 4 3 - 3 8 6	(5) 6 2 5 + 2 6 9
(6) 7 4 9 - 4 8 6	(7) 1 8 6 + 5 4 3	(8) 9 1 7 - 2 9 8	(9) 6 8 7 + 1 6 2	(10) 9 3 2 - 2 4 9
(11) 1 4 3 + 6 5 4	(12) 7 3 9 - 2 2 7	(13) 1 8 4 + 6 1 2	(14) 6 2 7 - 2 7 4	(15) 5 9 4 + 1 5 7
(16) 8 7 6 - 2 1 7	(17) 2 1 9 + 6 7 5	(18) 8 6 4 - 3 4 5	(19) 1 9 7 + 5 4 1	(20) 9 7 8 - 3 2 7
(21) 7 3 2 + 1 4 8	(22) 8 1 6 - 2 8 9	(23) 5 8 9 + 1 5 3	(24) 9 4 3 - 2 5 6	(25) 6 2 5 + 1 7 9
(26) 8 4 9 - 2 8 7	(27) 5 4 3 + 2 7 6	(28) 9 1 7 - 2 9 8	(29) 1 5 2 + 6 8 7	(30) 9 3 2 - 2 1 7

Date:

Name:

Time:

Score: .... / 30

Addition & Subtraction (3 Digit)

①
```
   2 5 6
 + 4 2 1
 --------
  . . .
```

②
```
   8 3 2
 - 2 6 7
 --------
  . . .
```

③
```
   4 7 3
 + 3 6 2
 --------
  . . .
```

④
```
   7 4 5
 - 2 7 8
 --------
  . . .
```

⑤
```
   5 3 4
 + 1 4 6
 --------
  . . .
```

⑥
```
   9 2 6
 - 1 8 5
 --------
  . . .
```

⑦
```
   2 3 1
 + 5 1 2
 --------
  . . .
```

⑧
```
   7 8 9
 - 3 1 2
 --------
  . . .
```

⑨
```
   2 4 7
 + 6 2 3
 --------
  . . .
```

⑩
```
   5 2 8
 - 1 4 8
 --------
  . . .
```

⑪
```
   4 3 6
 + 2 1 4
 --------
  . . .
```

⑫
```
   8 5 7
 - 2 6 9
 --------
  . . .
```

⑬
```
   3 4 7
 + 5 2 1
 --------
  . . .
```

⑭
```
   7 8 3
 - 1 2 4
 --------
  . . .
```

⑮
```
   1 9 2
 + 4 2 8
 --------
  . . .
```

⑯
```
   7 3 4
 - 3 1 5
 --------
  . . .
```

⑰
```
   3 2 7
 + 5 9 2
 --------
  . . .
```

⑱
```
   8 7 5
 - 5 4 3
 --------
  . . .
```

⑲
```
   1 7 8
 + 4 6 1
 --------
  . . .
```

⑳
```
   8 4 5
 - 4 1 7
 --------
  . . .
```

㉑
```
   5 4 8
 + 3 4 5
 --------
  . . .
```

㉒
```
   7 5 6
 - 3 8 2
 --------
  . . .
```

㉓
```
   3 9 1
 + 2 8 6
 --------
  . . .
```

㉔
```
   8 7 2
 - 2 1 8
 --------
  . . .
```

㉕
```
   3 5 4
 + 5 6 4
 --------
  . . .
```

㉖
```
   9 3 1
 - 4 8 2
 --------
  . . .
```

㉗
```
   6 5 2
 + 2 3 8
 --------
  . . .
```

㉘
```
   7 8 5
 - 3 5 6
 --------
  . . .
```

㉙
```
   2 8 7
 + 4 9 8
 --------
  . . .
```

㉚
```
   6 2 7
 - 1 4 6
 --------
  . . .
```

Date: ....

Time: ....

Name: ....

Score: .... / 30

Addition & Subtraction (3 Digit)

1	2	3	4	5
5 7 2 + 3 1 4	8 1 6 - 2 8 5	4 2 7 + 2 6 3	6 7 4 - 2 9 5	2 1 5 + 5 4 2

6	7	8	9	10
8 2 1 - 3 8 7	4 3 7 + 2 5 9	6 7 4 - 2 3 5	5 7 6 + 3 1 6	8 3 5 - 4 1 7

11	12	13	14	15
6 2 1 + 2 7 1	8 2 4 - 3 8 5	5 3 7 + 3 1 6	8 2 6 - 3 8 4	6 5 9 + 2 4 3

16	17	18	19	20
7 4 5 - 3 5 6	5 6 4 + 3 1 9	7 4 8 - 4 2 3	2 2 8 + 6 4 5	9 1 5 - 3 8 4

21	22	23	24	25
6 7 2 + 3 1 4	8 7 2 - 4 3 7	5 9 6 + 3 1 7	9 2 7 - 4 9 8	7 1 3 + 1 9 4

26	27	28	29	30
9 4 5 - 5 2 7	6 2 7 + 2 8 1	8 3 5 - 4 6 8	5 2 9 + 3 3 1	8 7 6 - 4 5 6

Date: ....

Time: ....

Name: ....

Score: .... / 30

Addition & Subtraction (3 Digit)

① 634 + 268 · · ·	② 786 - 352 · · ·	③ 514 + 227 · · ·	④ 756 - 362 · · ·	⑤ 453 + 251 · · ·
⑥ 693 - 232 · · ·	⑦ 367 + 245 · · ·	⑧ 751 - 315 · · ·	⑨ 439 + 214 · · ·	⑩ 825 - 437 · · ·
⑪ 586 + 315 · · ·	⑫ 715 - 326 · · ·	⑬ 428 + 169 · · ·	⑭ 765 - 382 · · ·	⑮ 591 + 238 · · ·
⑯ 837 - 398 · · ·	⑰ 448 + 241 · · ·	⑱ 916 - 568 · · ·	⑲ 618 + 194 · · ·	⑳ 845 - 367 · · ·
㉑ 543 + 271 · · ·	㉒ 815 - 372 · · ·	㉓ 497 + 231 · · ·	㉔ 738 - 341 · · ·	㉕ 582 + 198 · · ·
㉖ 927 - 568 · · ·	㉗ 633 + 245 · · ·	㉘ 785 - 358 · · ·	㉙ 319 + 236 · · ·	㉚ 738 - 384 · · ·

Addition & Subtraction (3 Digit)

1)
$$452 + 218$$

2)
$$692 - 374$$

3)
$$715 + 163$$

4)
$$817 - 436$$

5)
$$527 + 192$$

6)
$$839 - 498$$

7)
$$582 + 271$$

8)
$$928 - 485$$

9)
$$613 + 289$$

10)
$$856 - 364$$

11)
$$567 + 227$$

12)
$$742 - 317$$

13)
$$486 + 216$$

14)
$$715 - 396$$

15)
$$534 + 431$$

16)
$$736 - 487$$

17)
$$623 + 171$$

18)
$$927 - 416$$

19)
$$574 + 319$$

20)
$$819 - 436$$

21)
$$642 + 286$$

22)
$$758 - 369$$

23)
$$573 + 218$$

24)
$$926 - 458$$

25)
$$649 + 315$$

26)
$$835 - 498$$

27)
$$539 + 271$$

28)
$$839 - 549$$

29)
$$638 + 320$$

30)
$$847 - 472$$

Date: ....

Time: ....

Name: ....

Score: .... / 30

Addition & Subtraction (3 Digit)

① 674 + 317 . . .	② 846 - 427 . . .	③ 532 + 271 . . .	④ 736 - 368 . . .	⑤ 458 + 214 . . .
⑥ 692 - 378 . . .	⑦ 537 + 129 . . .	⑧ 726 - 398 . . .	⑨ 548 + 271 . . .	⑩ 925 - 458 . . .
⑪ 657 + 229 . . .	⑫ 828 - 498 . . .	⑬ 617 + 271 . . .	⑭ 916 - 458 . . .	⑮ 639 + 338 . . .
⑯ 825 - 498 . . .	⑰ 271 + 512 . . .	⑱ 817 - 358 . . .	⑲ 315 + 534 . . .	⑳ 738 - 398 . . .
㉑ 271 + 549 . . .	㉒ 827 - 498 . . .	㉓ 239 + 561 . . .	㉔ 839 - 547 . . .	㉕ 638 + 271 . . .
㉖ 654 - 138 . . .	㉗ 423 + 269 . . .	㉘ 928 - 357 . . .	㉙ 543 + 239 . . .	㉚ 856 - 237 . . .

Date: ....

Name: ....

Time: ....

Score: .... / 30

Addition & Subtraction (3 Digit)

1	2	3	4	5
372 + 286	917 - 182	548 + 327	872 - 283	429 + 274

6	7	8	9	10
715 - 286	321 + 457	738 - 249	593 + 316	832 - 387

11	12	13	14	15
421 + 358	785 - 246	632 + 217	945 - 269	428 + 417

16	17	18	19	20
864 - 348	687 + 169	935 - 327	547 + 218	914 - 386

21	22	23	24	25
312 + 287	858 - 327	642 + 186	928 - 317	569 + 138

26	27	28	29	30
786 - 327	159 + 623	917 - 329	451 + 158	969 - 328

Date: ....

Time: ....

Name: ....

Score: .... / 30

Addition & Subtraction (3 Digit)

①
```
    4 5 6
+   2 3 4
─────────
  .   .   .
```

②
```
    8 9 9
−   4 8 5
─────────
  .   .   .
```

③
```
    3 1 2
+   5 7 8
─────────
  .   .   .
```

④
```
    8 4 3
−   2 1 9
─────────
  .   .   .
```

⑤
```
    5 4 3
+   3 4 2
─────────
  .   .   .
```

⑥
```
    7 6 6
−   3 4 7
─────────
  .   .   .
```

⑦
```
    6 5 4
+   1 2 6
─────────
  .   .   .
```

⑧
```
    9 3 7
−   3 4 3
─────────
  .   .   .
```

⑨
```
    2 8 8
+   5 1 9
─────────
  .   .   .
```

⑩
```
    7 2 8
−   2 6 8
─────────
  .   .   .
```

⑪
```
    5 2 3
+   4 6 7
─────────
  .   .   .
```

⑫
```
    8 1 7
−   2 3 5
─────────
  .   .   .
```

⑬
```
    3 6 5
+   2 7 4
─────────
  .   .   .
```

⑭
```
    6 7 8
−   4 3 5
─────────
  .   .   .
```

⑮
```
    7 3 5
+   1 8 9
─────────
  .   .   .
```

⑯
```
    9 5 7
−   3 8 7
─────────
  .   .   .
```

⑰
```
    2 4 4
+   3 6 7
─────────
  .   .   .
```

⑱
```
    8 3 3
−   2 9 8
─────────
  .   .   .
```

⑲
```
    4 4 1
+   1 4 7
─────────
  .   .   .
```

⑳
```
    9 7 2
−   3 1 4
─────────
  .   .   .
```

㉑
```
    3 8 9
+   5 1 2
─────────
  .   .   .
```

㉒
```
    7 5 7
−   2 4 5
─────────
  .   .   .
```

㉓
```
    4 5 9
+   1 5 4
─────────
  .   .   .
```

㉔
```
    6 9 1
−   2 1 9
─────────
  .   .   .
```

㉕
```
    5 3 7
+   2 4 3
─────────
  .   .   .
```

㉖
```
    8 5 5
−   3 1 9
─────────
  .   .   .
```

㉗
```
    6 8 3
+   1 6 4
─────────
  .   .   .
```

㉘
```
    9 2 7
−   2 4 5
─────────
  .   .   .
```

㉙
```
    3 6 6
+   5 3 1
─────────
  .   .   .
```

㉚
```
    7 6 8
−   2 8 5
─────────
  .   .   .
```

Date:

Name:

Time:

Score: .... / 30

Addition & Subtraction (3 Digit)

① 488 + 249	② 812 - 224	③ 332 + 477	④ 685 - 265	⑤ 412 + 329
⑥ 819 - 245	⑦ 288 + 468	⑧ 754 - 245	⑨ 527 + 198	⑩ 861 - 298
⑪ 355 + 517	⑫ 778 - 266	⑬ 475 + 154	⑭ 843 - 328	⑮ 466 + 199
⑯ 912 - 354	⑰ 534 + 244	⑱ 756 - 219	⑲ 288 + 312	⑳ 614 - 276
㉑ 236 + 366	㉒ 819 - 357	㉓ 476 + 374	㉔ 768 - 289	㉕ 389 + 278
㉖ 847 - 375	㉗ 576 + 184	㉘ 836 - 357	㉙ 288 + 287	㉚ 729 - 394

Addition & Subtraction (3 Digit)

1	2	3	4	5
457	861	244	799	315
+ 264	- 289	+ 167	- 244	+ 594
. . .	. . .	. . .	. . .	. . .

6	7	8	9	10
643	376	912	578	747
- 255	+ 485	- 446	+ 395	- 189
. . .	. . .	. . .	. . .	. . .

11	12	13	14	15
285	679	511	826	357
+ 288	- 398	+ 269	- 484	+ 279
. . .	. . .	. . .	. . .	. . .

16	17	18	19	20
683	487	917	366	736
- 267	+ 178	- 415	+ 324	- 244
. . .	. . .	. . .	. . .	. . .

21	22	23	24	25
399	827	377	835	355
+ 278	- 349	+ 245	- 367	+ 326
. . .	. . .	. . .	. . .	. . .

26	27	28	29	30
666	366	727	355	578
- 357	+ 489	- 356	+ 129	- 245
. . .	. . .	. . .	. . .	. . .

Date:

Time:

Name:

Score: _____ / 30

Addition & Subtraction (3 Digit)

1.
```
   3 4 7
 + 4 1 3
 _____
  . . .
```

2.
```
   9 6 3
 - 4 7 5
 _____
  . . .
```

3.
```
   4 1 2
 + 1 8 9
 _____
  . . .
```

4.
```
   7 4 6
 - 1 2 5
 _____
  . . .
```

5.
```
   5 4 3
 + 2 8 4
 _____
  . . .
```

6.
```
   9 1 7
 - 3 8 4
 _____
  . . .
```

7.
```
   6 9 8
 + 2 4 2
 _____
  . . .
```

8.
```
   8 3 4
 - 5 4 3
 _____
  . . .
```

9.
```
   3 2 3
 + 4 3 3
 _____
  . . .
```

10.
```
   7 7 5
 - 1 8 7
 _____
  . . .
```

11.
```
   4 7 5
 + 3 5 9
 _____
  . . .
```

12.
```
   8 1 2
 - 2 5 4
 _____
  . . .
```

13.
```
   5 8 6
 + 2 3 5
 _____
  . . .
```

14.
```
   9 1 6
 - 3 2 5
 _____
  . . .
```

15.
```
   3 2 9
 + 4 6 5
 _____
  . . .
```

16.
```
   4 3 7
 - 3 1 4
 _____
  . . .
```

17.
```
   5 2 7
 + 2 9 4
 _____
  . . .
```

18.
```
   7 6 5
 - 5 2 6
 _____
  . . .
```

19.
```
   4 6 7
 + 4 2 9
 _____
  . . .
```

20.
```
   7 5 4
 - 3 4 6
 _____
  . . .
```

21.
```
   4 9 8
 + 1 8 2
 _____
  . . .
```

22.
```
   6 6 9
 - 2 3 5
 _____
  . . .
```

23.
```
   5 2 9
 + 3 4 8
 _____
  . . .
```

24.
```
   9 6 4
 - 4 2 7
 _____
  . . .
```

25.
```
   6 4 1
 + 1 9 8
 _____
  . . .
```

26.
```
   9 2 5
 - 3 1 6
 _____
  . . .
```

27.
```
   5 4 7
 + 3 5 2
 _____
  . . .
```

28.
```
   8 9 7
 - 4 8 3
 _____
  . . .
```

29.
```
   5 7 4
 + 2 8 9
 _____
  . . .
```

30.
```
   8 4 3
 - 3 4 2
 _____
  . . .
```

Date:

Time:

Name:

Score: _____ / 30

Addition & Subtraction (3 Digit)

1)
```
   4 6 4
 + 4 5 8
---------
```

2)
```
   7 4 8
 - 2 8 5
---------
```

3)
```
   5 8 7
 + 2 2 9
---------
```

4)
```
   8 4 1
 - 3 5 9
---------
```

5)
```
   6 1 2
 + 3 7 7
---------
```

6)
```
   9 4 6
 - 4 1 5
---------
```

7)
```
   5 1 3
 + 3 8 5
---------
```

8)
```
   7 2 8
 - 2 7 4
---------
```

9)
```
   6 3 2
 + 2 5 9
---------
```

10)
```
   8 6 6
 - 3 8 9
---------
```

11)
```
   4 9 8
 + 2 8 3
---------
```

12)
```
   7 3 1
 - 3 6 9
---------
```

13)
```
   6 5 2
 + 3 2 4
---------
```

14)
```
   8 5 7
 - 4 6 1
---------
```

15)
```
   3 6 5
 + 3 2 9
---------
```

16)
```
   8 1 4
 - 4 7 4
---------
```

17)
```
   4 4 6
 + 3 5 5
---------
```

18)
```
   9 7 6
 - 4 6 3
---------
```

19)
```
   5 8 7
 + 2 5 4
---------
```

20)
```
   8 4 3
 - 4 2 4
---------
```

21)
```
   5 8 7
 + 2 3 6
---------
```

22)
```
   8 2 7
 - 1 6 0
---------
```

23)
```
   2 1 8
 + 4 5 7
---------
```

24)
```
   8 6 1
 - 2 8 3
---------
```

25)
```
   6 4 7
 + 2 4 8
---------
```

26)
```
   7 7 7
 - 4 8 6
---------
```

27)
```
   4 2 9
 + 2 6 8
---------
```

28)
```
   7 2 5
 - 3 1 7
---------
```

29)
```
   4 3 5
 + 3 5 7
---------
```

30)
```
   6 5 4
 - 4 4 8
---------
```

Date: ....

Time: ....

Name: ....

Score: .... / 30

Addition & Subtraction (3 Digit)

(1) 467 + 285	(2) 839 - 358	(3) 426 + 354	(4) 647 - 428	(5) 369 + 249
(6) 754 - 335	(7) 497 + 204	(8) 835 - 329	(9) 329 + 349	(10) 964 - 427
(11) 578 + 136	(12) 836 - 229	(13) 513 + 248	(14) 729 - 285	(15) 435 + 498
(16) 983 - 264	(17) 543 + 257	(18) 835 - 487	(19) 456 + 389	(20) 756 - 318
(21) 169 + 755	(22) 956 - 629	(23) 529 + 238	(24) 795 - 398	(25) 618 + 287
(26) 721 - 315	(27) 534 + 318	(28) 865 - 575	(29) 549 + 285	(30) 723 - 394

Addition & Subtraction (3 Digit)

1)
```
   2 4 5
 + 3 2 2
 ─────────
  .  .  .
```

2)
```
   9 8 8
 - 1 2 9
 ─────────
  .  .  .
```

3)
```
   4 8 7
 + 4 2 3
 ─────────
  .  .  .
```

4)
```
   8 8 7
 - 4 5 8
 ─────────
  .  .  .
```

5)
```
   3 1 9
 + 5 9 4
 ─────────
  .  .  .
```

6)
```
   9 5 5
 - 2 9 2
 ─────────
  .  .  .
```

7)
```
   5 4 8
 + 1 3 3
 ─────────
  .  .  .
```

8)
```
   9 6 7
 - 3 4 3
 ─────────
  .  .  .
```

9)
```
   4 2 7
 + 2 4 8
 ─────────
  .  .  .
```

10)
```
   7 2 6
 - 2 1 9
 ─────────
  .  .  .
```

11)
```
   5 6 7
 + 1 5 9
 ─────────
  .  .  .
```

12)
```
   9 8 9
 - 7 9 8
 ─────────
  .  .  .
```

13)
```
   3 2 9
 + 6 2 5
 ─────────
  .  .  .
```

14)
```
   8 5 4
 - 4 8 8
 ─────────
  .  .  .
```

15)
```
   6 3 6
 + 2 8 2
 ─────────
  .  .  .
```

16)
```
   9 2 7
 - 4 2 6
 ─────────
  .  .  .
```

17)
```
   4 8 3
 + 3 1 8
 ─────────
  .  .  .
```

18)
```
   9 5 5
 - 5 3 8
 ─────────
  .  .  .
```

19)
```
   1 9 2
 + 4 6 3
 ─────────
  .  .  .
```

20)
```
   8 6 7
 - 5 3 9
 ─────────
  .  .  .
```

21)
```
   1 3 7
 + 5 6 3
 ─────────
  .  .  .
```

22)
```
   8 6 8
 - 4 3 9
 ─────────
  .  .  .
```

23)
```
   5 9 4
 + 3 1 5
 ─────────
  .  .  .
```

24)
```
   9 7 6
 - 5 4 2
 ─────────
  .  .  .
```

25)
```
   4 2 3
 + 3 8 5
 ─────────
  .  .  .
```

26)
```
   8 2 5
 - 3 6 2
 ─────────
  .  .  .
```

27)
```
   5 2 8
 + 2 6 8
 ─────────
  .  .  .
```

28)
```
   9 3 5
 - 5 1 2
 ─────────
  .  .  .
```

29)
```
   6 5 4
 + 1 8 6
 ─────────
  .  .  .
```

30)
```
   8 4 5
 - 5 2 9
 ─────────
  .  .  .
```

Solution

Day 1

(1) 5 + 1 = 6	(2) 3 + 4 = 7	(3) 2 + 1 = 3	(4) 1 + 4 = 5	(5) 7 + 0 = 7
(6) 7 + 2 = 9	(7) 6 + 1 = 7	(8) 5 + 3 = 8	(9) 0 + 2 = 2	(10) 1 + 3 = 3
(11) 1 + 1 = 2	(12) 2 + 1 = 3	(13) 4 + 2 = 6	(14) 0 + 5 = 5	(15) 1 + 7 = 8
(16) 2 + 3 = 5	(17) 4 + 4 = 8	(18) 7 + 2 = 9	(19) 2 + 6 = 8	(20) 4 + 6 = 10
(21) 4 + 1 = 5	(22) 2 + 0 = 2	(23) 9 + 0 = 9	(24) 1 + 7 = 8	(25) 7 + 3 = 10
(26) 3 + 2 = 5	(27) 4 + 6 = 10	(28) 4 + 2 = 6	(29) 5 + 5 = 10	(30) 2 + 4 = 6

Day 2

(1) 1 + 7 = 8	(2) 3 + 5 = 8	(3) 4 + 3 = 7	(4) 8 + 4 = 12	(5) 5 + 2 = 7
(6) 6 + 1 = 7	(7) 2 + 7 = 9	(8) 5 + 6 = 11	(9) 10 + 0 = 10	(10) 9 + 5 = 14
(11) 6 + 2 = 8	(12) 3 + 8 = 11	(13) 6 + 6 = 12	(14) 8 + 2 = 10	(15) 3 + 4 = 7
(16) 3 + 9 = 12	(17) 6 + 6 = 12	(18) 5 + 7 = 12	(19) 9 + 2 = 11	(20) 6 + 3 = 9
(21) 9 + 3 = 12	(22) 5 + 9 = 14	(23) 8 + 1 = 9	(24) 4 + 9 = 13	(25) 8 + 4 = 12
(26) 7 + 7 = 14	(27) 5 + 3 = 8	(28) 2 + 9 = 11	(29) 1 + 0 = 1	(30) 5 + 2 = 7

Day 3

(1) 9 + 6 = 15	(2) 8 + 7 = 15	(3) 3 + 9 = 12	(4) 8 + 5 = 13	(5) 6 + 6 = 12
(6) 6 + 7 = 13	(7) 8 + 7 = 15	(8) 10 + 4 = 14	(9) 8 + 10 = 18	(10) 9 + 6 = 15
(11) 2 + 7 = 9	(12) 10 + 6 = 16	(13) 9 + 3 = 12	(14) 4 + 6 = 10	(15) 9 + 8 = 17
(16) 7 + 2 = 9	(17) 5 + 5 = 15	(18) 6 + 7 = 13	(19) 0 + 9 = 9	(20) 10 + 1 = 11
(21) 7 + 8 = 15	(22) 1 + 10 = 11	(23) 3 + 1 = 4	(24) 9 + 5 = 14	(25) 7 + 3 = 10
(26) 9 + 9 = 18	(27) 2 + 9 = 11	(28) 4 + 6 = 10	(29) 10 + 7 = 17	(30) 8 + 8 = 16

Day 4

(1) 7 + 5 = 12	(2) 5 + 9 = 14	(3) 4 + 7 = 11	(4) 2 + 8 = 10	(5) 9 + 6 = 15
(6) 8 + 8 = 16	(7) 10 + 7 = 17	(8) 9 + 10 = 19	(9) 3 + 9 = 12	(10) 9 + 3 = 12
(11) 3 + 6 = 9	(12) 5 + 8 = 13	(13) 1 + 9 = 10	(14) 8 + 7 = 15	(15) 6 + 6 = 12
(16) 5 + 9 = 14	(17) 10 + 5 = 15	(18) 10 + 8 = 18	(19) 6 + 3 = 9	(20) 8 + 6 = 14
(21) 2 + 9 = 11	(22) 8 + 3 = 11	(23) 7 + 8 = 15	(24) 3 + 8 = 11	(25) 8 + 4 = 12
(26) 9 + 8 = 17	(27) 5 + 9 = 14	(28) 6 + 9 = 15	(29) 5 + 8 = 13	(30) 10 + 1 = 11

Solution

Day 5

(1) 10 + 10 = 20	(2) 8 + 4 = 12	(3) 6 + 9 = 15	(4) 9 + 6 = 15	(5) 8 + 9 = 17
(6) 7 + 8 = 15	(7) 9 + 9 = 18	(8) 10 + 2 = 12	(9) 8 + 5 = 13	(10) 6 + 8 = 14
(11) 8 + 3 = 11	(12) 9 + 4 = 13	(13) 10 + 8 = 18	(14) 2 + 10 = 12	(15) 8 + 8 = 16
(16) 8 + 8 = 16	(17) 4 + 9 = 13	(18) 5 + 7 = 12	(19) 5 + 6 = 11	(20) 7 + 7 = 14
(21) 10 + 0 = 11	(22) 5 + 8 = 13	(23) 7 + 3 = 10	(24) 7 + 9 = 16	(25) 8 + 7 = 15
(26) 4 + 9 = 13	(27) 7 + 5 = 12	(28) 6 + 7 = 13	(29) 3 + 8 = 11	(30) 10 + 3 = 13

Day 6

(1) 11 + 10 = 21	(2) 12 + 13 = 25	(3) 15 + 12 = 27	(4) 13 + 16 = 29	(5) 11 + 18 = 29
(6) 17 + 12 = 29	(7) 10 + 19 = 29	(8) 14 + 12 = 26	(9) 12 + 15 = 27	(10) 15 + 12 = 27
(11) 11 + 13 = 24	(12) 10 + 14 = 24	(13) 13 + 15 = 28	(14) 15 + 10 = 25	(15) 10 + 10 = 20
(16) 17 + 12 = 29	(17) 14 + 13 = 27	(18) 18 + 10 = 28	(19) 15 + 13 = 28	(20) 16 + 11 = 27
(21) 15 + 10 = 25	(22) 16 + 12 = 28	(23) 17 + 11 = 28	(24) 10 + 18 = 28	(25) 12 + 17 = 29
(26) 14 + 14 = 28	(27) 15 + 12 = 27	(28) 16 + 10 = 26	(29) 13 + 15 = 28	(30) 10 + 15 = 25

Day 7

(1) 13 + 16 = 29	(2) 11 + 11 = 22	(3) 12 + 12 = 24	(4) 15 + 10 = 25	(5) 16 + 12 = 28
(6) 11 + 17 = 28	(7) 14 + 13 = 27	(8) 15 + 11 = 26	(9) 10 + 17 = 27	(10) 17 + 12 = 29
(11) 18 + 10 = 28	(12) 13 + 17 = 30	(13) 19 + 12 = 31	(14) 11 + 17 = 28	(15) 14 + 15 = 29
(16) 18 + 13 = 31	(17) 16 + 10 = 26	(18) 14 + 14 = 28	(19) 19 + 14 = 33	(20) 11 + 15 = 26
(21) 13 + 12 = 25	(22) 17 + 17 = 34	(23) 14 + 10 = 24	(24) 17 + 12 = 29	(25) 12 + 18 = 30
(26) 12 + 19 = 31	(27) 13 + 15 = 28	(28) 16 + 11 = 27	(29) 14 + 17 = 31	(30) 17 + 10 = 27

Day 8

(1) 20 + 13 = 33	(2) 15 + 18 = 33	(3) 14 + 16 = 30	(4) 19 + 12 = 31	(5) 12 + 14 = 26
(6) 18 + 15 = 33	(7) 17 + 14 = 31	(8) 11 + 10 = 21	(9) 13 + 13 = 26	(10) 10 + 18 = 28
(11) 20 + 10 = 30	(12) 18 + 11 = 29	(13) 12 + 18 = 30	(14) 11 + 14 = 25	(15) 13 + 12 = 25
(16) 15 + 17 = 32	(17) 13 + 13 = 26	(18) 18 + 10 = 28	(19) 12 + 20 = 32	(20) 10 + 17 = 27
(21) 20 + 12 = 32	(22) 12 + 14 = 28	(23) 19 + 17 = 36	(24) 13 + 15 = 28	(25) 15 + 10 = 25
(26) 16 + 17 = 33	(27) 15 + 13 = 28	(28) 18 + 18 = 36	(29) 17 + 14 = 31	(30) 19 + 10 = 29

Solution

Day 9

① 10 + 19 = 29	② 20 + 16 = 36	③ 17 + 18 = 35	④ 10 + 16 = 26	⑤ 18 + 13 = 31
⑥ 16 + 19 = 35	⑦ 15 + 13 = 28	⑧ 13 + 20 = 33	⑨ 17 + 14 = 31	⑩ 20 + 18 = 38
⑪ 15 + 14 = 29	⑫ 19 + 10 = 29	⑬ 15 + 20 = 35	⑭ 14 + 13 = 27	⑮ 15 + 16 = 31
⑯ 17 + 20 = 37	⑰ 16 + 18 = 34	⑱ 19 + 11 = 30	⑲ 20 + 20 = 40	⑳ 18 + 10 = 28
㉑ 18 + 15 = 33	㉒ 17 + 13 = 30	㉓ 17 + 20 = 37	㉔ 14 + 18 = 32	㉕ 13 + 16 = 29
㉖ 14 + 14 = 28	㉗ 16 + 17 = 33	㉘ 15 + 17 = 32	㉙ 15 + 20 = 35	㉚ 19 + 11 = 30

Day 10

① 12 + 20 = 32	② 20 + 13 = 33	③ 14 + 18 = 32	④ 20 + 17 = 37	⑤ 19 + 12 = 31
⑥ 14 + 19 = 33	⑦ 10 + 17 = 27	⑧ 16 + 14 = 30	⑨ 18 + 15 = 33	⑩ 20 + 13 = 33
⑪ 12 + 17 = 29	⑫ 15 + 16 = 31	⑬ 18 + 20 = 38	⑭ 20 + 20 = 40	⑮ 15 + 17 = 32
⑯ 11 + 10 = 21	⑰ 16 + 20 = 36	⑱ 19 + 19 = 38	⑲ 20 + 15 = 35	⑳ 17 + 10 = 27
㉑ 18 + 14 = 32	㉒ 20 + 13 = 33	㉓ 12 + 10 = 22	㉔ 15 + 18 = 33	㉕ 14 + 16 = 30
㉖ 18 + 12 = 30	㉗ 20 + 15 = 35	㉘ 15 + 15 = 30	㉙ 15 + 19 = 34	㉚ 20 + 10 = 30

Day 11

① 1 − 0 = 1	② 2 − 0 = 2	③ 2 − 1 = 1	④ 3 − 1 = 2	⑤ 3 − 2 = 1
⑥ 4 − 2 = 2	⑦ 4 − 3 = 1	⑧ 5 − 3 = 2	⑨ 5 − 5 = 0	⑩ 6 − 4 = 2
⑪ 5 − 4 = 1	⑫ 6 − 5 = 1	⑬ 7 − 4 = 3	⑭ 8 − 3 = 5	⑮ 7 − 3 = 4
⑯ 7 − 6 = 1	⑰ 8 − 6 = 2	⑱ 9 − 7 = 2	⑲ 9 − 5 = 4	⑳ 4 − 0 = 4
㉑ 8 − 4 = 4	㉒ 4 − 3 = 1	㉓ 2 − 0 = 2	㉔ 6 − 3 = 3	㉕ 6 − 6 = 0
㉖ 8 − 2 = 6	㉗ 6 − 5 = 1	㉘ 9 − 8 = 1	㉙ 9 − 4 = 5	㉚ 10 − 9 = 1

Day 12

① 4 − 1 = 3	② 5 − 2 = 3	③ 6 − 2 = 4	④ 7 − 3 = 4	⑤ 8 − 5 = 3
⑥ 9 − 3 = 6	⑦ 10 − 5 = 5	⑧ 5 − 1 = 4	⑨ 7 − 2 = 5	⑩ 9 − 2 = 7
⑪ 8 − 2 = 6	⑫ 8 − 4 = 4	⑬ 9 − 8 = 1	⑭ 10 − 1 = 9	⑮ 8 − 8 = 0
⑯ 10 − 2 = 8	⑰ 9 − 1 = 8	⑱ 7 − 1 = 6	⑲ 6 − 3 = 3	⑳ 4 − 4 = 0
㉑ 8 − 5 = 3	㉒ 9 − 6 = 3	㉓ 10 − 6 = 4	㉔ 9 − 5 = 4	㉕ 8 − 3 = 5
㉖ 9 − 1 = 8	㉗ 10 − 9 = 1	㉘ 8 − 7 = 1	㉙ 7 − 2 = 5	㉚ 9 − 7 = 2

Solution

Day 13

#		#		#		#		#	
①	5 − 2 = 3	②	8 − 3 = 5	③	6 − 4 = 2	④	9 − 3 = 6	⑤	7 − 1 = 6
⑥	10 − 3 = 7	⑦	7 − 3 = 4	⑧	9 − 4 = 5	⑨	6 − 2 = 4	⑩	8 − 1 = 7
⑪	10 − 7 = 3	⑫	5 − 3 = 2	⑬	9 − 6 = 3	⑭	7 − 2 = 5	⑮	10 − 3 = 7
⑯	8 − 2 = 6	⑰	3 − 1 = 2	⑱	9 − 5 = 4	⑲	7 − 3 = 4	⑳	10 − 4 = 6
㉑	7 − 4 = 3	㉒	2 − 1 = 1	㉓	8 − 4 = 4	㉔	6 − 1 = 5	㉕	9 − 5 = 4
㉖	7 − 2 = 5	㉗	4 − 1 = 3	㉘	3 − 2 = 1	㉙	10 − 8 = 2	㉚	6 − 5 = 1

Day 14

#		#		#		#		#	
①	9 − 7 = 2	②	7 − 5 = 2	③	8 − 6 = 2	④	4 − 2 = 2	⑤	10 − 9 = 1
⑥	9 − 1 = 8	⑦	6 − 4 = 2	⑧	7 − 3 = 4	⑨	5 − 2 = 3	⑩	8 − 5 = 3
⑪	10 − 6 = 4	⑫	7 − 1 = 6	⑬	9 − 3 = 6	⑭	6 − 2 = 4	⑮	8 − 1 = 7
⑯	10 − 7 = 3	⑰	5 − 3 = 2	⑱	9 − 6 = 3	⑲	7 − 2 = 5	⑳	10 − 3 = 7
㉑	8 − 2 = 6	㉒	3 − 1 = 2	㉓	9 − 5 = 4	㉔	6 − 3 = 3	㉕	10 − 4 = 6
㉖	7 − 4 = 3	㉗	2 − 1 = 1	㉘	8 − 4 = 4	㉙	6 − 1 = 5	㉚	9 − 5 = 4

Day 15

#		#		#		#		#	
①	9 − 2 = 7	②	8 − 3 = 5	③	10 − 4 = 6	④	7 − 1 = 6	⑤	6 − 4 = 2
⑥	10 − 6 = 4	⑦	7 − 3 = 4	⑧	9 − 4 = 5	⑨	8 − 1 = 7	⑩	10 − 7 = 3
⑪	9 − 6 = 3	⑫	7 − 2 = 5	⑬	10 − 3 = 7	⑭	8 − 2 = 6	⑮	9 − 1 = 8
⑯	6 − 3 = 3	⑰	10 − 4 = 6	⑱	7 − 4 = 3	⑲	8 − 6 = 2	⑳	10 − 9 = 1
㉑	9 − 1 = 8	㉒	6 − 5 = 1	㉓	9 − 7 = 2	㉔	7 − 5 = 2	㉕	8 − 4 = 4
㉖	10 − 6 = 4	㉗	9 − 3 = 6	㉘	8 − 5 = 3	㉙	7 − 2 = 5	㉚	10 − 3 = 7

Day 16

#		#		#		#		#	
①	15 − 10 = 05	②	18 − 11 = 07	③	14 − 10 = 04	④	17 − 13 = 04	⑤	19 − 11 = 08
⑥	20 − 12 = 08	⑦	16 − 10 = 06	⑧	20 − 13 = 07	⑨	12 − 10 = 02	⑩	19 − 14 = 05
⑪	13 − 10 = 03	⑫	16 − 12 = 04	⑬	18 − 13 = 05	⑭	20 − 15 = 05	⑮	17 − 11 = 06
⑯	19 − 16 = 03	⑰	14 − 12 = 02	⑱	20 − 17 = 03	⑲	11 − 10 = 01	⑳	18 − 15 = 03
㉑	15 − 12 = 03	㉒	20 − 18 = 02	㉓	16 − 14 = 02	㉔	19 − 17 = 02	㉕	13 − 12 = 01
㉖	20 − 19 = 01	㉗	17 − 15 = 02	㉘	18 − 16 = 02	㉙	14 − 13 = 01	㉚	20 − 20 = 00

Solution

Day 17

(1) 18 − 11 = 07	(2) 19 − 13 = 06	(3) 17 − 12 = 05	(4) 20 − 15 = 05	(5) 16 − 11 = 05
(6) 14 − 10 = 04	(7) 20 − 16 = 04	(8) 19 − 14 = 05	(9) 18 − 12 = 06	(10) 17 − 10 = 07
(11) 16 − 14 = 02	(12) 13 − 10 = 03	(13) 20 − 17 = 03	(14) 15 − 12 = 03	(15) 19 − 15 = 04
(16) 18 − 13 = 05	(17) 17 − 11 = 06	(18) 16 − 12 = 04	(19) 20 − 18 = 02	(20) 14 − 13 = 01
(21) 19 − 16 = 03	(22) 15 − 10 = 05	(23) 18 − 14 = 04	(24) 17 − 15 = 02	(25) 20 − 19 = 01
(26) 16 − 13 = 03	(27) 19 − 17 = 02	(28) 18 − 15 = 03	(29) 17 − 13 = 04	(30) 16 − 10 = 06

Day 18

(1) 20 − 20 = 00	(2) 19 − 18 = 01	(3) 18 − 16 = 02	(4) 17 − 14 = 03	(5) 16 − 15 = 01
(6) 15 − 14 = 01	(7) 14 − 12 = 02	(8) 13 − 10 = 03	(9) 20 − 11 = 09	(10) 19 − 12 = 07
(11) 18 − 17 = 01	(12) 17 − 16 = 01	(13) 15 − 13 = 02	(14) 14 − 11 = 03	(15) 16 − 14 = 02
(16) 13 − 10 = 03	(17) 20 − 13 = 07	(18) 19 − 15 = 04	(19) 18 − 10 = 08	(20) 17 − 12 = 05
(21) 16 − 11 = 05	(22) 16 − 16 = 00	(23) 14 − 10 = 04	(24) 20 − 14 = 06	(25) 19 − 16 = 03
(26) 18 − 13 = 05	(27) 17 − 15 = 02	(28) 16 − 12 = 04	(29) 15 − 11 = 04	(30) 14 − 12 = 02

Day 19

(1) 19 − 10 = 09	(2) 20 − 11 = 09	(3) 18 − 12 = 06	(4) 17 − 13 = 04	(5) 16 − 10 = 06
(6) 20 − 12 = 08	(7) 20 − 12 = 08	(8) 19 − 13 = 06	(9) 18 − 17 = 01	(10) 19 − 10 = 09
(11) 16 − 11 = 05	(12) 20 − 14 = 06	(13) 19 − 15 = 04	(14) 18 − 10 = 08	(15) 17 − 12 = 05
(16) 16 − 13 = 03	(17) 15 − 14 = 01	(18) 20 − 16 = 04	(19) 19 − 17 = 02	(20) 18 − 16 = 02
(21) 16 − 15 = 01	(22) 15 − 12 = 03	(23) 20 − 17 = 03	(24) 19 − 18 = 01	(25) 19 − 16 = 03
(26) 17 − 15 = 02	(27) 16 − 14 = 02	(28) 15 − 13 = 02	(29) 20 − 19 = 01	(30) 18 − 11 = 07

Day 20

(1) 18 − 11 = 07	(2) 17 − 10 = 07	(3) 16 − 12 = 04	(4) 20 − 13 = 07	(5) 19 − 17 = 02
(6) 18 − 15 = 03	(7) 17 − 12 = 05	(8) 16 − 15 = 01	(9) 15 − 11 = 04	(10) 20 − 15 = 05
(11) 19 − 16 = 03	(12) 18 − 14 = 04	(13) 17 − 12 = 05	(14) 16 − 10 = 06	(15) 15 − 10 = 05
(16) 20 − 16 = 04	(17) 19 − 17 = 02	(18) 18 − 13 = 06	(19) 17 − 14 = 03	(20) 16 − 11 = 05
(21) 15 − 12 = 03	(22) 14 − 13 = 01	(23) 20 − 18 = 02	(24) 19 − 15 = 04	(25) 18 − 17 = 01
(26) 11 − 10 = 01	(27) 16 − 13 = 03	(28) 15 − 11 = 04	(29) 14 − 12 = 02	(30) 13 − 10 = 03

Solution

Day 21

#		#		#		#		#	
1	5 + 12 = 17	2	20 − 3 = 17	3	8 + 9 = 17	4	14 − 2 = 12	5	10 + 6 = 16
6	18 − 8 = 10	7	3 + 11 = 14	8	20 − 5 = 15	9	2 + 13 = 15	10	16 − 6 = 10
11	7 + 9 = 16	12	15 − 3 = 12	13	4 + 10 = 14	14	13 − 4 = 11	15	9 + 8 = 17
16	20 − 2 = 18	17	1 + 14 = 15	18	17 − 6 = 11	19	6 + 11 = 17	20	20 − 7 = 13
21	3 + 12 = 15	22	12 − 4 = 12	23	8 + 11 = 19	24	20 − 8 = 12	25	2 + 15 = 17
26	18 − 5 = 13	27	5 + 13 = 18	28	14 − 5 = 09	29	10 + 7 = 17	30	15 − 4 = 11

Day 22

#		#		#		#		#	
1	4 + 11 = 15	2	13 − 3 = 10	3	10 + 9 = 19	4	20 − 6 = 14	5	16 + 1 = 17
6	17 − 4 = 13	7	12 + 7 = 19	8	16 − 2 = 14	9	6 + 9 = 15	10	20 − 10 = 10
11	13 + 3 = 16	12	14 − 4 = 10	13	11 + 7 = 18	14	18 − 3 = 15	15	8 + 10 = 18
16	20 − 5 = 15	17	2 + 14 = 16	18	16 − 3 = 13	19	12 + 7 = 19	20	15 − 2 = 13
21	10 + 6 = 16	22	19 − 5 = 14	23	4 + 12 = 16	24	13 − 2 = 11	25	9 + 7 = 16
26	16 − 1 = 15	27	15 + 1 = 16	28	19 − 2 = 17	29	7 + 11 = 18	30	20 − 1 = 19

Day 23

#		#		#		#		#	
1	3 + 10 = 13	2	14 − 3 = 11	3	8 + 9 = 17	4	19 − 4 = 15	5	6 + 8 = 14
6	15 − 1 = 14	7	5 + 9 = 14	8	18 − 3 = 15	9	4 + 8 = 12	10	20 − 1 = 19
11	12 + 2 = 14	12	17 − 1 = 16	13	10 + 7 = 17	14	16 − 4 = 12	15	9 + 6 = 15
16	14 − 1 = 13	17	7 + 6 = 13	18	20 − 3 = 17	19	11 + 3 = 14	20	13 − 1 = 12
21	5 + 6 = 11	22	18 − 4 = 14	23	8 + 7 = 15	24	15 − 2 = 13	25	4 + 5 = 09
26	13 − 1 = 12	27	6 + 4 = 10	28	20 − 4 = 16	29	11 + 2 = 13	30	12 − 1 = 11

Day 24

#		#		#		#		#	
1	12 + 4 = 16	2	17 − 7 = 10	3	7 + 8 = 15	4	14 − 6 = 08	5	10 + 7 = 17
6	19 − 9 = 10	7	10 + 3 = 13	8	17 − 5 = 12	9	14 + 2 = 16	10	16 − 5 = 11
11	9 + 7 = 16	12	15 − 3 = 12	13	11 + 4 = 15	14	13 − 4 = 09	15	9 + 9 = 18
16	20 − 2 = 18	17	15 + 1 = 16	18	17 − 6 = 11	19	12 + 6 = 18	20	20 − 8 = 12
21	13 + 3 = 16	22	12 − 3 = 09	23	10 + 8 = 18	24	19 − 5 = 14	25	16 + 2 = 18
26	18 − 4 = 14	27	12 + 5 = 17	28	14 − 4 = 10	29	10 + 8 = 18	30	15 − 2 = 13

Solution

Day 25

(1) 10 + 5 = 15	(2) 13 − 3 = 10	(3) 11 + 9 = 20	(4) 20 − 7 = 13	(5) 16 + 1 = 17
(6) 17 − 4 = 13	(7) 13 + 7 = 20	(8) 16 − 2 = 14	(9) 10 + 6 = 16	(10) 20 − 10 = 10
(11) 12 + 3 = 15	(12) 14 − 5 = 09	(13) 11 + 7 = 18	(14) 18 − 3 = 15	(15) 11 + 8 = 19
(16) 20 − 6 = 14	(17) 15 + 2 = 17	(18) 17 − 3 = 14	(19) 13 + 5 = 18	(20) 15 − 1 = 14
(21) 10 + 6 = 16	(22) 19 − 5 = 14	(23) 12 + 4 = 16	(24) 13 − 2 = 11	(25) 9 + 8 = 17
(26) 16 − 1 = 15	(27) 18 + 2 = 20	(28) 17 − 3 = 14	(29) 11 + 7 = 18	(30) 20 − 5 = 15

Day 26

(1) 13 + 5 = 18	(2) 10 − 4 = 06	(3) 12 + 4 = 16	(4) 9 − 3 = 6	(5) 4 + 3 = 7
(6) 11 − 6 = 05	(7) 11 + 2 = 13	(8) 7 − 1 = 6	(9) 6 + 1 = 7	(10) 13 − 8 = 05
(11) 2 + 1 = 3	(12) 9 − 5 = 4	(13) 4 + 2 = 6	(14) 12 − 8 = 04	(15) 3 + 1 = 4
(16) 8 − 2 = 6	(17) 7 + 3 = 10	(18) 15 − 9 = 06	(19) 1 + 4 = 5	(20) 10 − 5 = 05
(21) 5 + 2 = 7	(22) 11 − 7 = 04	(23) 2 + 3 = 5	(24) 8 − 4 = 4	(25) 16 + 4 = 20
(26) 13 − 9 = 04	(27) 12 + 3 = 15	(28) 19 − 5 = 14	(29) 14 + 1 = 15	(30) 12 − 7 = 05

Day 27

(1) 13 + 2 = 15	(2) 17 − 2 = 15	(3) 10 + 5 = 15	(4) 15 − 8 = 07	(5) 12 + 4 = 16
(6) 12 − 5 = 07	(7) 10 + 1 = 11	(8) 15 − 1 = 14	(9) 14 + 6 = 20	(10) 13 − 4 = 09
(11) 20 + 5 = 25	(12) 18 − 3 = 15	(13) 17 + 2 = 19	(14) 14 − 6 = 08	(15) 14 + 3 = 17
(16) 12 − 3 = 09	(17) 10 + 3 = 13	(18) 16 − 4 = 12	(19) 15 + 6 = 21	(20) 12 − 12 = 00
(21) 16 + 2 = 18	(22) 13 − 5 = 08	(23) 20 + 10 = 30	(24) 18 − 2 = 16	(25) 15 + 3 = 18
(26) 12 − 4 = 08	(27) 10 + 7 = 17	(28) 19 − 2 = 17	(29) 17 + 1 = 18	(30) 14 − 7 = 07

Day 28

(1) 3 + 13 = 16	(2) 10 − 1 = 09	(3) 16 + 4 = 20	(4) 13 − 6 = 07	(5) 11 + 8 = 19
(6) 18 − 1 = 17	(7) 15 + 4 = 19	(8) 11 − 3 = 08	(9) 12 + 8 = 20	(10) 19 − 1 = 18
(11) 20 + 4 = 24	(12) 15 − 7 = 08	(13) 13 + 6 = 19	(14) 12 − 3 = 09	(15) 11 + 9 = 20
(16) 10 − 10 = 00	(17) 14 + 5 = 19	(18) 14 − 5 = 09	(19) 12 + 9 = 21	(20) 18 − 9 = 09
(21) 16 + 9 = 25	(22) 20 − 10 = 10	(23) 16 + 5 = 21	(24) 13 − 4 = 09	(25) 11 + 10 = 21
(26) 19 − 9 = 10	(27) 15 + 13 = 28	(28) 20 − 1 = 19	(29) 12 + 8 = 20	(30) 16 − 5 = 11

Solution

Day 29

#	Problem		#	Problem		#	Problem		#	Problem		#	Problem
1	$13 + 15 = 28$		2	$19 - 4 = 15$		3	$13 + 4 = 17$		4	$10 - 6 = 04$		5	$15 + 3 = 18$
6	$12 - 6 = 06$		7	$11 + 6 = 17$		8	$17 - 4 = 13$		9	$16 + 2 = 18$		10	$13 - 8 = 05$
11	$2 + 11 = 13$		12	$19 - 6 = 13$		13	$14 + 2 = 16$		14	$11 - 8 = 03$		15	$13 + 3 = 16$
16	$18 - 2 = 16$		17	$20 + 5 = 25$		18	$15 - 9 = 06$		19	$11 + 4 = 15$		20	$10 - 5 = 05$
21	$15 + 5 = 20$		22	$12 - 7 = 05$		23	$13 + 2 = 15$		24	$18 - 6 = 12$		25	$20 + 7 = 27$
26	$13 - 6 = 07$		27	$11 + 9 = 20$		28	$19 - 5 = 14$		29	$13 + 8 = 21$		30	$11 - 7 = 04$

Day 30

#	Problem		#	Problem		#	Problem		#	Problem		#	Problem
1	$12 + 9 = 21$		2	$18 - 12 = 06$		3	$16 + 5 = 21$		4	$13 - 4 = 09$		5	$11 + 11 = 22$
6	$12 - 3 = 09$		7	$11 + 9 = 20$		8	$10 - 1 = 09$		9	$14 + 5 = 19$		10	$14 - 12 = 02$
11	$12 + 8 = 20$		12	$19 - 2 = 17$		13	$17 + 4 = 21$		14	$15 - 7 = 08$		15	$13 + 6 = 19$
16	$13 - 6 = 07$		17	$11 + 8 = 19$		18	$18 - 18 = 00$		19	$20 + 9 = 29$		20	$11 - 3 = 08$
21	$17 + 1 = 18$		22	$14 - 7 = 07$		23	$13 + 3 = 16$		24	$10 - 2 = 08$		25	$16 + 3 = 19$
26	$18 - 2 = 16$		27	$15 + 10 = 25$		28	$12 - 10 = 02$		29	$12 + 7 = 19$		30	$19 - 7 = 12$

Day 31

#	Problem		#	Problem		#	Problem		#	Problem		#	Problem
1	$18 - 3 = 15$		2	$14 + 7 = 21$		3	$19 - 6 = 13$		4	$12 + 3 = 15$		5	$15 - 12 = 03$
6	$15 + 4 = 19$		7	$16 - 1 = 15$		8	$11 + 2 = 13$		9	$17 - 5 = 12$		10	$10 + 5 = 15$
11	$9 + 6 = 15$		12	$4 - 2 = 2$		13	$16 + 2 = 18$		14	$5 - 3 = 2$		15	$12 + 5 = 17$
16	$8 - 2 = 6$		17	$10 + 3 = 13$		18	$6 - 4 = 2$		19	$14 + 1 = 15$		20	$11 - 5 = 06$
21	$13 + 4 = 17$		22	$9 - 2 = 7$		23	$15 + 2 = 17$		24	$7 - 1 = 6$		25	$17 + 2 = 19$
26	$14 - 1 = 13$		27	$12 + 6 = 18$		28	$6 - 2 = 4$		29	$10 + 4 = 14$		30	$8 - 5 = 3$

Day 32

#	Problem		#	Problem		#	Problem		#	Problem		#	Problem
1	$14 + 5 = 19$		2	$5 - 1 = 4$		3	$11 + 1 = 12$		4	$13 - 2 = 11$		5	$14 + 3 = 17$
6	$19 - 3 = 16$		7	$15 + 4 = 19$		8	$17 - 2 = 15$		9	$17 + 3 = 20$		10	$12 - 11 = 01$
11	$18 + 2 = 20$		12	$12 - 2 = 10$		13	$13 + 5 = 18$		14	$14 - 1 = 13$		15	$11 + 3 = 14$
16	$13 - 5 = 08$		17	$19 + 9 = 28$		18	$15 - 2 = 13$		19	$14 + 12 = 26$		20	$16 - 2 = 14$
21	$12 + 4 = 16$		22	$18 - 4 = 14$		23	$10 + 1 = 11$		24	$16 - 3 = 13$		25	$14 + 6 = 20$
26	$11 - 2 = 09$		27	$20 + 7 = 27$		28	$19 - 8 = 11$		29	$19 + 1 = 20$		30	$16 - 4 = 12$

Solution

Day 33

#	Problem	Answer
1	4 + 7	11
2	9 − 6	3
3	12 + 3	15
4	5 − 2	3
5	15 + 4	19
6	6 − 1	5
7	11 + 2	13
8	7 − 4	3
9	10 + 5	15
10	3 − 1	2
11	13 + 6	19
12	2 − 1	1
13	14 + 3	17
14	7 − 3	4
15	9 + 6	15
16	4 − 2	2
17	16 + 2	18
18	5 − 3	2
19	12 + 5	17
20	8 − 2	6
21	10 + 3	13
22	16 − 4	12
23	14 + 1	15
24	11 − 5	06
25	13 + 4	17
26	9 − 2	7
27	15 + 2	17
28	16 − 4	12
29	15 + 3	18
30	4 − 1	3

Day 34

#	Problem	Answer
1	12 + 6	18
2	9 − 2	7
3	10 + 4	14
4	8 − 5	3
5	14 + 5	19
6	5 − 1	4
7	11 + 1	12
8	3 − 2	1
9	13 + 3	16
10	9 − 3	6
11	15 + 1	16
12	17 − 2	15
13	16 + 4	20
14	12 − 4	08
15	20 + 5	25
16	10 − 1	09
17	17 + 2	19
18	14 − 1	13
19	11 + 3	08
20	13 − 5	08
21	19 + 1	20
22	15 − 3	12
23	17 + 2	19
24	16 − 2	14
25	12 + 2	14
26	18 − 4	14
27	10 + 1	11
28	16 − 3	13
29	14 + 2	16
30	11 − 2	09

Day 35

#	Problem	Answer
1	13 + 5	18
2	19 − 1	18
3	15 + 3	18
4	17 − 4	13
5	16 + 6	22
6	12 − 5	13
7	8 + 6	14
8	18 − 9	09
9	5 + 9	14
10	16 − 3	13
11	10 + 3	13
12	19 − 7	12
13	6 + 8	14
14	15 − 7	08
15	9 + 2	11
16	17 − 4	13
17	7 + 5	12
18	20 − 6	14
19	12 + 4	16
20	14 − 2	12
21	8 + 3	11
22	19 − 6	13
23	15 + 2	17
24	16 − 8	08
25	10 + 2	12
26	17 − 6	11
27	6 + 5	11
28	15 − 3	12
29	19 + 6	25
30	17 − 8	09

Day 36

#	Problem	Answer
1	7 + 4	11
2	20 − 9	11
3	12 + 3	15
4	14 − 3	11
5	8 + 5	13
6	19 − 5	14
7	5 + 7	12
8	16 − 4	12
9	10 + 4	14
10	18 − 6	12
11	6 + 9	15
12	15 − 8	07
13	9 + 3	12
14	17 − 6	11
15	7 + 3	10
16	20 − 7	13
17	12 + 6	18
18	14 − 1	13
19	8 + 4	12
20	19 − 9	10
21	5 + 8	13
22	16 − 2	14
23	10 + 5	15
24	18 − 5	13
25	6 + 7	13
26	15 − 5	10
27	9 + 5	14
28	17 − 7	10
29	7 + 6	13
30	20 − 8	12

Solution

Day 37

#		#		#		#		#	
①	12 + 2 = 14	②	14 − 4 = 10	③	8 + 2 = 10	④	19 − 4 = 15	⑤	5 + 3 = 8
⑥	16 − 7 = 09	⑦	10 + 1 = 11	⑧	18 − 8 = 10	⑨	6 + 4 = 10	⑩	15 − 3 = 12
⑪	9 + 1 = 10	⑫	17 − 3 = 14	⑬	7 + 2 = 9	⑭	20 − 6 = 14	⑮	12 + 5 = 17
⑯	14 − 6 = 08	⑰	8 + 1 = 9	⑱	19 − 3 = 16	⑲	5 + 6 = 11	⑳	16 − 6 = 10
㉑	10 + 6 = 16	㉒	18 − 2 = 16	㉓	6 + 3 = 9	㉔	15 − 8 = 07	㉕	9 + 4 = 13
㉖	17 − 5 = 12	㉗	7 + 1 = 8	㉘	20 − 5 = 15	㉙	12 + 1 = 13	㉚	14 − 5 = 09

Day 38

#		#		#		#		#	
①	7 + 4 = 11	②	12 − 6 = 06	③	5 + 2 = 7	④	18 − 9 = 09	⑤	10 + 3 = 13
⑥	14 − 8 = 06	⑦	13 + 6 = 19	⑧	20 − 10 = 10	⑨	8 + 5 = 13	⑩	16 − 7 = 09
⑪	6 + 11 = 17	⑫	13 − 4 = 09	⑬	4 + 15 = 19	⑭	17 − 9 = 08	⑮	9 + 2 = 11
⑯	15 − 6 = 09	⑰	2 + 8 = 10	⑱	19 − 11 = 08	⑲	11 + 4 = 15	⑳	20 − 8 = 12
㉑	7 + 12 = 19	㉒	12 − 5 = 07	㉓	5 + 14 = 19	㉔	18 − 7 = 11	㉕	10 + 1 = 11
㉖	14 − 6 = 08	㉗	13 + 3 = 16	㉘	19 − 7 = 12	㉙	18 + 11 = 29	㉚	16 − 4 = 12

Day 39

#		#		#		#		#	
①	16 + 12 = 28	②	13 − 3 = 10	③	14 + 6 = 20	④	17 − 8 = 09	⑤	19 + 1 = 20
⑥	15 − 5 = 10	⑦	12 + 7 = 19	⑧	19 − 10 = 09	⑨	11 + 3 = 14	⑩	20 − 9 = 11
⑪	7 + 13 = 20	⑫	12 − 1 = 11	⑬	15 + 4 = 19	⑭	18 − 6 = 12	⑮	10 + 2 = 12
⑯	14 − 7 = 07	⑰	3 + 14 = 17	⑱	20 − 11 = 09	⑲	18 + 4 = 22	⑳	16 − 6 = 10
㉑	6 + 3 = 9	㉒	13 − 2 = 11	㉓	4 + 17 = 21	㉔	17 − 7 = 10	㉕	9 + 3 = 12
㉖	15 − 4 = 11	㉗	12 + 9 = 21	㉘	19 − 8 = 11	㉙	11 + 2 = 13	㉚	20 − 7 = 13

Day 40

#		#		#		#		#	
①	7 + 11 = 18	②	12 − 4 = 08	③	5 + 6 = 11	④	18 − 5 = 13	⑤	10 + 4 = 06
⑥	14 − 9 = 05	⑦	4 + 15 = 19	⑧	20 − 6 = 14	⑨	8 + 2 = 10	⑩	16 − 4 = 12
⑪	6 + 4 = 10	⑫	13 − 5 = 12	⑬	4 + 8 = 12	⑭	17 − 5 = 12	⑮	9 + 4 = 13
⑯	15 − 3 = 12	⑰	2 + 16 = 18	⑱	19 − 9 = 10	⑲	11 + 1 = 12	⑳	20 − 5 = 15
㉑	7 + 5 = 12	㉒	12 − 2 = 10	㉓	5 + 7 = 12	㉔	18 − 4 = 14	㉕	10 + 5 = 15
㉖	14 − 3 = 11	㉗	13 + 2 = 15	㉘	20 − 4 = 16	㉙	18 + 1 = 19	㉚	16 − 3 = 19

Solution

Day 41

1) 11 + 3 = 14
2) 14 − 4 = 10
3) 9 + 6 = 15
4) 19 − 4 = 15
5) 8 + 4 = 12
6) 15 − 1 = 14
7) 9 + 5 = 14
8) 18 − 2 = 16
9) 8 + 6 = 14
10) 20 − 3 = 17
11) 13 + 4 = 17
12) 17 − 1 = 16
13) 10 + 7 = 17
14) 16 − 4 = 12
15) 9 + 9 = 18
16) 14 − 1 = 13
17) 6 + 5 = 11
18) 18 − 4 = 14
19) 8 + 5 = 13
20) 15 − 2 = 13
21) 4 + 4 = 8
22) 13 − 1 = 12
23) 6 + 4 = 10
24) 20 − 20 = 00
25) 12 + 2 = 14
26) 20 − 1 = 19
27) 12 + 1 = 13
28) 18 − 8 = 10
29) 10 + 9 = 19
30) 15 − 5 = 10

Day 42

1) 18 + 4 = 22
2) 20 − 7 = 13
3) 12 + 9 = 21
4) 19 − 9 = 10
5) 14 + 6 = 20
6) 20 − 8 = 12
7) 15 + 10 = 25
8) 17 − 5 = 12
9) 19 + 3 = 22
10) 20 − 11 = 09
11) 17 + 7 = 24
12) 20 − 6 = 14
13) 12 + 14 = 26
14) 20 − 4 = 16
15) 16 + 5 = 21
16) 19 − 2 = 17
17) 18 + 2 = 20
18) 20 − 9 = 11
19) 20 + 9 = 29
20) 19 − 8 = 11
21) 16 + 4 = 20
22) 18 − 3 = 15
23) 19 + 10 = 09
24) 20 − 7 = 13
25) 18 + 7 = 25
26) 17 − 4 = 13
27) 19 + 6 = 25
28) 20 − 5 = 15
29) 13 + 17 = 30
30) 20 − 1 = 19

Day 43

1) 19 + 11 = 30
2) 20 − 2 = 18
3) 19 + 8 = 27
4) 20 − 7 = 13
5) 20 + 1 = 21
6) 20 − 9 = 11
7) 15 + 5 = 20
8) 20 − 3 = 17
9) 18 + 13 = 31
10) 20 − 10 = 10
11) 17 + 3 = 20
12) 20 − 12 = 08
13) 19 + 14 = 33
14) 19 − 4 = 15
15) 13 + 8 = 05
16) 20 − 5 = 15
17) 20 + 12 = 32
18) 20 − 7 = 13
19) 14 + 4 = 18
20) 20 − 6 = 14
21) 20 + 8 = 28
22) 19 − 3 = 16
23) 16 + 6 = 22
24) 10 − 2 = 08
25) 20 + 11 = 31
26) 20 − 15 = 05
27) 17 + 4 = 21
28) 19 − 11 = 08
29) 15 + 9 = 24
30) 20 − 11 = 09

Day 44

1) 13 + 20 = 33
2) 20 − 8 = 12
3) 16 + 7 = 23
4) 20 − 2 = 18
5) 20 + 10 = 30
6) 20 − 6 = 14
7) 19 + 4 = 23
8) 20 − 5 = 15
9) 15 + 6 = 21
10) 20 − 4 = 16
11) 12 + 18 = 30
12) 20 − 6 = 14
13) 17 + 8 = 25
14) 10 − 5 = 05
15) 14 + 5 = 19
16) 20 − 4 = 16
17) 12 + 9 = 21
18) 20 − 2 = 18
19) 20 + 3 = 23
20) 18 − 8 = 10
21) 20 + 14 = 34
22) 20 − 14 = 06
23) 12 + 7 = 19
24) 20 − 1 = 19
25) 15 + 11 = 26
26) 20 − 3 = 17
27) 13 + 4 = 17
28) 20 − 2 = 18
29) 12 + 6 = 18
30) 20 − 1 = 19

Solution

Day 45

(1) 17 + 4 = 21	(2) 20 − 6 = 14	(3) 13 + 7 = 20	(4) 18 − 9 = 09	(5) 15 + 7 = 22
(6) 19 − 7 = 12	(7) 16 + 11 = 27	(8) 20 − 5 = 15	(9) 18 + 3 = 21	(10) 20 − 10 = 10
(11) 15 + 8 = 23	(12) 20 − 7 = 13	(13) 17 + 12 = 29	(14) 20 − 3 = 17	(15) 14 + 6 = 20
(16) 20 − 2 = 18	(17) 19 + 2 = 21	(18) 20 − 8 = 12	(19) 20 + 10 = 30	(20) 20 − 9 = 11
(21) 4 + 15 = 19	(22) 20 − 4 = 16	(23) 19 + 7 = 26	(24) 20 − 6 = 14	(25) 16 + 13 = 29
(26) 20 − 4 = 16	(27) 18 + 5 = 23	(28) 20 − 7 = 13	(29) 19 + 9 = 28	(30) 20 − 1 = 19

Day 46

(1) 11 + 18 = 29	(2) 20 − 3 = 17	(3) 17 + 8 = 25	(4) 20 − 5 = 15	(5) 20 + 1 = 21
(6) 20 − 8 = 12	(7) 16 + 16 = 32	(8) 20 − 3 = 17	(9) 19 + 10 = 29	(10) 20 − 9 = 11
(11) 17 + 3 = 20	(12) 20 − 15 = 05	(13) 14 + 19 = 33	(14) 20 − 4 = 16	(15) 13 + 7 = 20
(16) 16 − 5 = 11	(17) 20 + 12 = 32	(18) 20 − 7 = 13	(19) 14 + 4 = 18	(20) 20 − 6 = 14
(21) 8 + 20 = 28	(22) 19 − 3 = 16	(23) 16 + 6 = 22	(24) 20 − 2 = 18	(25) 11 + 20 = 31
(26) 19 − 5 = 14	(27) 20 + 5 = 25	(28) 20 − 14 = 06	(29) 15 + 9 = 24	(30) 18 − 3 = 15

Day 47

(1) 13 + 20 = 33	(2) 20 − 8 = 12	(3) 16 + 7 = 23	(4) 20 − 2 = 18	(5) 20 + 10 = 30
(6) 20 − 6 = 14	(7) 19 + 4 = 23	(8) 20 − 5 = 15	(9) 15 + 6 = 21	(10) 18 − 8 = 10
(11) 19 + 12 = 31	(12) 20 − 6 = 14	(13) 17 + 8 = 25	(14) 17 − 7 = 10	(15) 14 + 5 = 19
(16) 13 − 4 = 09	(17) 12 + 9 = 21	(18) 18 − 9 = 09	(19) 20 + 3 = 23	(20) 17 − 6 = 11
(21) 14 + 20 = 34	(22) 19 − 9 = 10	(23) 12 + 6 = 28	(24) 20 − 1 = 19	(25) 13 + 24 = 37
(26) 19 − 6 = 13	(27) 18 + 6 = 24	(28) 16 − 5 = 11	(29) 17 + 13 = 30	(30) 20 − 7 = 13

Day 48

(1) 19 − 8 = 11	(2) 11 + 18 = 29	(3) 9 + 10 = 19	(4) 13 − 4 = 09	(5) 18 − 7 = 11
(6) 7 + 9 = 16	(7) 15 − 3 = 12	(8) 19 + 12 = 31	(9) 17 − 5 = 12	(10) 10 + 14 = 24
(11) 16 − 6 = 10	(12) 6 + 7 = 13	(13) 19 − 9 = 10	(14) 14 + 16 = 30	(15) 10 + 5 = 15
(16) 18 − 6 = 12	(17) 13 + 8 = 21	(18) 17 − 4 = 13	(19) 11 + 15 = 26	(20) 19 − 7 = 12
(21) 9 + 12 = 21	(22) 16 − 5 = 11	(23) 17 + 13 = 30	(24) 11 + 7 = 18	(25) 19 − 6 = 13
(26) 10 + 13 = 23	(27) 18 − 4 = 14	(28) 14 + 15 = 29	(29) 10 + 8 = 18	(30) 20 − 9 = 11

Solution

Day 49

(1) 12 + 14 = 26	(2) 17 − 3 = 14	(3) 9 + 6 = 15	(4) 16 − 7 = 09	(5) 13 + 15 = 28
(6) 9 + 8 = 17	(7) 19 − 4 = 15	(8) 11 + 13 = 24	(9) 18 − 3 = 15	(10) 11 + 10 = 21
(11) 15 − 6 = 11	(12) 7 + 8 = 15	(13) 19 − 3 = 16	(14) 14 + 12 = 26	(15) 16 − 4 = 12
(16) 13 + 14 = 27	(17) 6 + 6 = 12	(18) 19 − 2 = 17	(19) 10 + 9 = 19	(20) 18 − 5 = 13
(21) 12 + 11 = 23	(22) 15 − 7 = 08	(23) 7 + 8 = 15	(24) 19 − 1 = 18	(25) 12 + 12 = 24
(26) 17 − 6 = 11	(27) 9 + 6 = 15	(28) 14 + 9 = 23	(29) 19 − 5 = 14	(30) 13 − 12 = 01

Day 50

(1) 16 − 3 = 13	(2) 7 + 5 = 12	(3) 19 − 11 = 08	(4) 10 + 8 = 18	(5) 18 − 7 = 11
(6) 11 + 10 = 21	(7) 15 − 5 = 10	(8) 8 + 6 = 14	(9) 19 − 2 = 17	(10) 12 + 10 = 22
(11) 17 − 4 = 13	(12) 9 + 5 = 14	(13) 14 + 8 = 22	(14) 19 − 1 = 18	(15) 13 + 11 = 24
(16) 16 − 6 = 10	(17) 7 + 4 = 11	(18) 18 − 3 = 15	(19) 11 + 9 = 20	(20) 15 − 4 = 11
(21) 8 + 5 = 13	(22) 19 − 1 = 18	(23) 10 + 7 = 17	(24) 17 − 5 = 12	(25) 12 + 9 = 21
(26) 14 + 7 = 21	(27) 19 − 4 = 15	(28) 13 + 10 = 23	(29) 16 − 2 = 14	(30) 7 + 3 = 10

Day 51

(1) 15 + 25 = 40	(2) 45 − 10 = 35	(3) 20 + 31 = 51	(4) 51 − 20 = 31	(5) 35 + 10 = 45
(6) 55 − 10 = 45	(7) 25 + 22 = 47	(8) 60 − 30 = 30	(9) 32 + 11 = 43	(10) 43 − 10 = 33
(11) 46 + 13 = 59	(12) 65 − 20 = 45	(13) 10 + 41 = 51	(14) 50 − 30 = 20	(15) 21 + 21 = 42
(16) 72 − 30 = 42	(17) 25 + 25 = 50	(18) 88 − 20 = 68	(19) 34 + 15 = 49	(20) 55 − 20 = 35
(21) 15 + 43 = 58	(22) 72 − 41 = 31	(23) 25 + 10 = 35	(24) 55 − 25 = 30	(25) 23 + 30 = 53
(26) 60 − 40 = 20	(27) 30 + 12 = 42	(28) 54 − 24 = 30	(29) 10 + 33 = 43	(30) 36 − 13 = 23

Day 52

(1) 33 + 15 = 48	(2) 77 − 30 = 47	(3) 28 + 21 = 49	(4) 60 − 20 = 40	(5) 25 + 23 = 48
(6) 80 − 20 = 60	(7) 35 + 11 = 46	(8) 63 − 40 = 23	(9) 13 + 15 = 28	(10) 99 − 10 = 89
(11) 72 + 10 = 82	(12) 35 − 15 = 20	(13) 30 + 18 = 48	(14) 55 − 24 = 31	(15) 31 + 44 = 75
(16) 70 − 40 = 30	(17) 28 + 11 = 39	(18) 53 − 12 = 41	(19) 14 + 15 = 29	(20) 36 − 30 = 06
(21) 65 + 33 = 98	(22) 95 − 74 = 21	(23) 43 + 21 = 64	(24) 78 − 15 = 63	(25) 10 + 79 = 89
(26) 42 − 31 = 11	(27) 80 + 11 = 91	(28) 76 − 45 = 31	(29) 18 + 61 = 79	(30) 78 − 55 = 23

Solution

Day 53

(1) 42 + 23 = 65	(2) 89 − 32 = 57	(3) 11 + 31 = 42	(4) 68 − 37 = 31	(5) 51 + 17 = 68
(6) 97 − 43 = 54	(7) 19 + 35 = 54	(8) 76 − 34 = 42	(9) 27 + 41 = 68	(10) 86 − 23 = 63
(11) 13 + 19 = 32	(12) 69 − 26 = 43	(13) 35 + 42 = 77	(14) 95 − 44 = 51	(15) 23 + 36 = 59
(16) 58 − 25 = 33	(17) 17 + 16 = 33	(18) 67 − 23 = 44	(19) 14 + 28 = 42	(20) 78 − 47 = 31
(21) 29 + 38 = 67	(22) 85 − 41 = 44	(23) 33 + 19 = 52	(24) 76 − 22 = 54	(25) 26 + 45 = 71
(26) 97 − 52 = 45	(27) 12 + 36 = 48	(28) 75 − 35 = 40	(29) 21 + 31 = 52	(30) 65 − 35 = 30

Day 54

(1) 18 + 45 = 63	(2) 87 − 24 = 63	(3) 36 + 15 = 51	(4) 56 − 35 = 21	(5) 16 + 18 = 34
(6) 47 − 26 = 21	(7) 21 + 24 = 45	(8) 68 − 32 = 36	(9) 23 + 16 = 39	(10) 79 − 48 = 31
(11) 15 + 15 = 30	(12) 68 − 15 = 53	(13) 28 + 41 = 69	(14) 97 − 31 = 66	(15) 37 + 13 = 50
(16) 75 − 24 = 51	(17) 14 + 28 = 42	(18) 58 − 24 = 34	(19) 26 + 48 = 74	(20) 89 − 56 = 33
(21) 13 + 12 = 25	(22) 64 − 33 = 31	(23) 32 + 28 = 60	(24) 68 − 14 = 54	(25) 19 + 14 = 33
(26) 49 − 17 = 32	(27) 17 + 48 = 65	(28) 79 − 26 = 53	(29) 31 + 68 = 99	(30) 99 − 39 = 60

Day 55

(1) 22 + 13 = 35	(2) 88 − 23 = 65	(3) 26 + 45 = 71	(4) 89 − 55 = 34	(5) 19 + 24 = 43
(6) 57 − 34 = 23	(7) 11 + 28 = 39	(8) 48 − 16 = 32	(9) 33 + 17 = 50	(10) 78 − 32 = 46
(11) 13 + 55 = 68	(12) 77 − 32 = 45	(13) 25 + 45 = 70	(14) 86 − 22 = 64	(15) 16 + 43 = 59
(16) 99 − 59 = 40	(17) 12 + 19 = 31	(18) 59 − 27 = 32	(19) 18 + 17 = 35	(20) 49 − 28 = 21
(21) 23 + 24 = 47	(22) 78 − 36 = 42	(23) 17 + 31 = 48	(24) 69 − 17 = 52	(25) 29 + 16 = 45
(26) 86 − 42 = 44	(27) 32 + 17 = 49	(28) 59 − 33 = 26	(29) 24 + 29 = 53	(30) 76 − 24 = 52

Day 56

(1) 14 + 32 = 46	(2) 62 − 55 = 07	(3) 31 + 26 = 57	(4) 78 − 37 = 41	(5) 15 + 23 = 38
(6) 49 − 16 = 33	(7) 18 + 46 = 64	(8) 95 − 49 = 46	(9) 11 + 28 = 39	(10) 59 − 38 = 21
(11) 27 + 35 = 62	(12) 85 − 39 = 46	(13) 12 + 28 = 40	(14) 77 − 45 = 32	(15) 28 + 19 = 47
(16) 64 − 28 = 36	(17) 19 + 21 = 40	(18) 78 − 39 = 39	(19) 25 + 46 = 71	(20) 93 − 55 = 38
(21) 18 + 14 = 32	(22) 49 − 18 = 31	(23) 15 + 19 = 34	(24) 65 − 26 = 39	(25) 23 + 24 = 47
(26) 77 − 43 = 34	(27) 27 + 39 = 66	(28) 81 − 38 = 43	(29) 22 + 14 = 36	(30) 68 − 32 = 36

Solution

Day 57

(1) 34 + 46 = 80	(2) 95 − 45 = 50	(3) 19 + 17 = 36	(4) 54 − 28 = 26	(5) 18 + 31 = 49
(6) 69 − 35 = 34	(7) 25 + 38 = 63	(8) 92 − 55 = 37	(9) 14 + 19 = 33	(10) 75 − 27 = 48
(11) 29 + 45 = 74	(12) 96 − 38 = 58	(13) 22 + 14 = 36	(14) 58 − 26 = 32	(15) 18 + 38 = 46
(16) 77 − 35 = 42	(17) 23 + 24 = 47	(18) 67 − 49 = 18	(19) 16 + 19 = 35	(20) 54 − 29 = 25
(21) 27 + 15 = 42	(22) 66 − 28 = 38	(23) 17 + 66 = 83	(24) 58 − 28 = 30	(25) 43 + 35 = 78
(26) 72 − 38 = 34	(27) 16 + 49 = 65	(28) 97 − 42 = 55	(29) 18 + 17 = 35	(30) 63 − 29 = 34

Day 58

(1) 29 + 16 = 45	(2) 78 − 39 = 39	(3) 13 + 25 = 38	(4) 64 − 26 = 38	(5) 27 + 21 = 48
(6) 71 − 25 = 46	(7) 21 + 16 = 37	(8) 48 − 29 = 19	(9) 18 + 35 = 53	(10) 86 − 37 = 49
(11) 23 + 28 = 51	(12) 59 − 35 = 24	(13) 16 + 42 = 58	(14) 84 − 45 = 39	(15) 12 + 16 = 28
(16) 47 − 23 = 24	(17) 22 + 37 = 59	(18) 79 − 38 = 41	(19) 14 + 31 = 45	(20) 66 − 35 = 31
(21) 24 + 17 = 41	(22) 85 − 46 = 39	(23) 16 + 31 = 47	(24) 88 − 37 = 51	(25) 23 + 38 = 61
(26) 73 − 35 = 38	(27) 18 + 45 = 63	(28) 86 − 38 = 48	(29) 13 + 19 = 32	(30) 47 − 28 = 19

Day 59

(1) 25 + 27 = 52	(2) 68 − 38 = 30	(3) 18 + 15 = 33	(4) 43 − 24 = 19	(5) 14 + 25 = 39
(6) 52 − 26 = 26	(7) 19 + 32 = 51	(8) 65 − 19 = 46	(9) 29 + 28 = 57	(10) 71 − 45 = 26
(11) 43 + 19 = 62	(12) 68 − 28 = 40	(13) 22 + 38 = 60	(14) 75 − 36 = 39	(15) 14 + 19 = 33
(16) 58 − 25 = 33	(17) 15 + 47 = 62	(18) 84 − 24 = 60	(19) 27 + 18 = 45	(20) 59 − 25 = 34
(21) 21 + 28 = 49	(22) 76 − 39 = 37	(23) 19 + 37 = 56	(24) 91 − 18 = 73	(25) 26 + 14 = 40
(26) 42 − 27 = 15	(27) 13 + 36 = 49	(28) 56 − 24 = 32	(29) 21 + 24 = 45	(30) 71 − 29 = 42

Day 60

(1) 27 + 37 = 64	(2) 85 − 39 = 46	(3) 12 + 24 = 36	(4) 64 − 35 = 29	(5) 19 + 17 = 36
(6) 43 − 26 = 17	(7) 17 + 28 = 45	(8) 69 − 38 = 31	(9) 24 + 27 = 51	(10) 87 − 42 = 45
(11) 16 + 15 = 31	(12) 53 − 27 = 26	(13) 25 + 18 = 43	(14) 79 − 48 = 31	(15) 13 + 28 = 41
(16) 72 − 28 = 44	(17) 28 + 41 = 69	(18) 85 − 38 = 47	(19) 18 + 17 = 35	(20) 67 − 28 = 39
(21) 31 + 18 = 49	(22) 69 − 35 = 34	(23) 18 + 28 = 46	(24) 62 − 35 = 27	(25) 24 + 18 = 42
(26) 91 − 35 = 56	(27) 16 + 31 = 47	(28) 53 − 29 = 24	(29) 13 + 16 = 29	(30) 48 − 26 = 22

Solution

Day 61

#		#		#		#		#	
①	12 + 19 = 31	②	43 − 24 = 19	③	14 + 36 = 50	④	69 − 35 = 34	⑤	27 + 24 = 51
⑥	87 − 39 = 48	⑦	13 + 15 = 28	⑧	54 − 29 = 25	⑨	23 + 38 = 61	⑩	75 − 38 = 37
⑪	17 + 35 = 52	⑫	69 − 32 = 37	⑬	26 + 29 = 55	⑭	76 − 39 = 37	⑮	13 + 21 = 34
⑯	48 − 23 = 25	⑰	16 + 28 = 44	⑱	62 − 27 = 35	⑲	19 + 18 = 37	⑳	46 − 28 = 18
㉑	27 + 19 = 46	㉒	85 − 35 = 50	㉓	14 + 25 = 39	㉔	59 − 28 = 31	㉕	23 + 38 = 61
㉖	71 − 38 = 33	㉗	16 + 39 = 55	㉘	78 − 32 = 46	㉙	19 + 15 = 34	㉚	43 − 23 = 20

Day 62

#		#		#		#		#	
①	56 + 23 = 79	②	87 − 34 = 53	③	43 + 21 = 64	④	75 − 16 = 59	⑤	66 + 13 = 79
⑥	95 − 22 = 73	⑦	48 + 37 = 85	⑧	82 − 41 = 41	⑨	54 + 12 = 66	⑩	68 − 23 = 45
⑪	32 + 45 = 77	⑫	99 − 38 = 61	⑬	59 + 28 = 87	⑭	74 − 19 = 55	⑮	41 + 34 = 75
⑯	88 − 43 = 45	⑰	63 + 11 = 74	⑱	91 − 16 = 75	⑲	52 + 17 = 69	⑳	77 − 35 = 42
㉑	36 + 42 = 78	㉒	98 − 17 = 81	㉓	62 + 19 = 81	㉔	83 − 25 = 58	㉕	57 + 15 = 72
㉖	92 − 31 = 61	㉗	45 + 28 = 73	㉘	79 − 24 = 55	㉙	61 + 18 = 79	㉚	84 − 22 = 62

Day 63

#		#		#		#		#	
①	33 + 27 = 60	②	95 − 63 = 32	③	46 + 14 = 60	④	76 − 29 = 47	⑤	38 + 33 = 71
⑥	89 − 42 = 47	⑦	64 + 16 = 80	⑧	93 − 17 = 76	⑨	58 + 18 = 76	⑩	72 − 43 = 29
⑪	51 + 22 = 73	⑫	85 − 39 = 46	⑬	47 + 26 = 73	⑭	71 − 27 = 44	⑮	39 + 32 = 71
⑯	97 − 64 = 33	⑰	53 + 16 = 69	⑱	88 − 36 = 52	⑲	66 + 19 = 85	⑳	94 − 29 = 65
㉑	34 + 47 = 81	㉒	76 − 42 = 34	㉓	49 + 23 = 72	㉔	69 − 28 = 41	㉕	42 + 36 = 78
㉖	90 − 44 = 46	㉗	67 + 18 = 85	㉘	96 − 22 = 74	㉙	55 + 12 = 67	㉚	80 − 38 = 42

Day 64

#		#		#		#		#	
①	35 + 14 = 49	②	74 − 48 = 26	③	56 + 17 = 73	④	81 − 12 = 69	⑤	43 + 23 = 66
⑥	98 − 62 = 36	⑦	59 + 16 = 75	⑧	86 − 24 = 62	⑨	63 + 14 = 77	⑩	92 − 37 = 55
⑪	31 + 45 = 76	⑫	73 − 26 = 47	⑬	57 + 13 = 70	⑭	78 − 24 = 54	⑮	40 + 36 = 76
⑯	87 − 49 = 38	⑰	64 + 19 = 83	⑱	93 − 51 = 42	⑲	48 + 15 = 63	⑳	82 − 29 = 53
㉑	37 + 16 = 53	㉒	75 − 41 = 34	㉓	54 + 13 = 67	㉔	79 − 28 = 51	㉕	44 + 27 = 71
㉖	96 − 65 = 31	㉗	61 + 15 = 76	㉘	89 − 47 = 42	㉙	65 + 11 = 76	㉚	91 − 19 = 72

Solution

Day 65

#	Problem	#	Problem	#	Problem	#	Problem	#	Problem
1	21 + 27 = 48	2	68 − 26 = 42	3	17 + 19 = 36	4	39 − 26 = 13	5	18 + 29 = 47
6	56 − 26 = 30	7	23 + 17 = 40	8	71 − 38 = 33	9	19 + 28 = 47	10	79 − 42 = 37
11	12 + 24 = 36	12	59 − 28 = 31	13	16 + 29 = 45	14	68 − 38 = 30	15	23 + 14 = 37
16	48 − 23 = 25	17	19 + 28 = 47	18	73 − 29 = 44	19	27 + 17 = 44	20	66 − 29 = 37
21	17 + 19 = 36	22	45 − 28 = 17	23	15 + 37 = 52	24	57 − 25 = 32	25	22 + 28 = 50
26	75 − 37 = 38	27	15 + 26 = 41	28	64 − 35 = 29	29	19 + 27 = 46	30	77 − 45 = 32

Day 66

#	Problem	#	Problem	#	Problem	#	Problem	#	Problem
1	32 + 36 = 68	2	83 − 41 = 42	3	58 + 17 = 75	4	70 − 26 = 44	5	45 + 29 = 74
6	77 − 35 = 42	7	62 + 18 = 80	8	94 − 45 = 49	9	57 + 14 = 71	10	88 − 32 = 56
11	45 + 23 = 68	12	78 − 36 = 42	13	14 + 39 = 53	14	91 − 18 = 73	15	56 + 27 = 83
16	85 − 44 = 41	17	32 + 58 = 90	18	97 − 22 = 75	19	76 + 15 = 91	20	83 − 59 = 24
21	20 + 58 = 78	22	77 − 12 = 65	23	61 + 36 = 97	24	92 − 25 = 67	25	43 + 54 = 97
26	88 − 11 = 77	27	59 + 31 = 90	28	66 − 22 = 44	29	39 + 26 = 65	30	84 − 42 = 42

Day 67

#	Problem	#	Problem	#	Problem	#	Problem	#	Problem
1	17 + 58 = 75	2	95 − 27 = 68	3	51 + 44 = 95	4	89 − 28 = 61	5	70 + 26 = 96
6	74 − 18 = 56	7	33 + 43 = 76	8	93 − 21 = 72	9	22 + 67 = 89	10	81 − 29 = 52
11	62 + 37 = 99	12	86 − 48 = 38	13	48 + 42 = 90	14	74 − 31 = 43	15	53 + 45 = 98
16	89 − 33 = 56	17	37 + 26 = 63	18	92 − 16 = 76	19	67 + 15 = 82	20	81 − 55 = 26
21	55 + 23 = 78	22	83 − 28 = 55	23	38 + 45 = 83	24	97 − 12 = 85	25	64 + 21 = 85
26	75 − 42 = 33	27	26 + 43 = 69	28	94 − 25 = 69	29	16 + 67 = 83	30	82 − 33 = 49

Day 68

#	Problem	#	Problem	#	Problem	#	Problem	#	Problem
1	33 + 64 = 97	2	77 − 13 = 64	3	42 + 36 = 78	4	68 − 27 = 41	5	19 + 38 = 57
6	85 − 45 = 40	7	54 + 14 = 68	8	91 − 39 = 52	9	29 + 55 = 84	10	78 − 26 = 52
11	46 + 23 = 69	12	95 − 58 = 37	13	39 + 24 = 63	14	72 − 28 = 44	15	54 + 42 = 96
16	86 − 19 = 67	17	25 + 59 = 84	18	83 − 47 = 36	19	49 + 26 = 75	20	76 − 14 = 62
21	34 + 45 = 79	22	92 − 35 = 57	23	48 + 18 = 66	24	62 − 27 = 35	25	25 + 55 = 80
26	87 − 13 = 74	27	35 + 44 = 89	28	96 − 32 = 64	29	14 + 64 = 78	30	84 − 49 = 35

Solution

Day 69

(1) 54 + 25 = 79	(2) 73 − 21 = 52	(3) 28 + 49 = 77	(4) 97 − 65 = 32	(5) 17 + 38 = 55
(6) 62 − 28 = 34	(7) 45 + 36 = 81	(8) 85 − 41 = 44	(9) 32 + 49 = 81	(10) 93 − 22 = 71
(11) 57 + 34 = 91	(12) 79 − 43 = 36	(13) 39 + 35 = 74	(14) 65 − 27 = 38	(15) 18 + 76 = 94
(16) 81 − 34 = 47	(17) 56 + 23 = 79	(18) 94 − 28 = 66	(19) 31 + 68 = 99	(20) 88 − 46 = 42
(21) 39 + 17 = 56	(22) 82 − 42 = 40	(23) 76 + 12 = 88	(24) 68 − 22 = 46	(25) 27 + 38 = 65
(26) 95 − 13 = 82	(27) 24 + 63 = 87	(28) 41 − 15 = 26	(29) 84 + 11 = 95	(30) 72 − 48 = 24

Day 70

(1) 35 + 42 = 77	(2) 78 − 25 = 53	(3) 57 + 22 = 79	(4) 86 − 43 = 43	(5) 62 + 18 = 80
(6) 91 − 27 = 64	(7) 40 + 15 = 55	(8) 72 − 30 = 42	(9) 46 + 23 = 69	(10) 88 − 16 = 72
(11) 24 + 33 = 57	(12) 95 − 42 = 53	(13) 53 + 16 = 69	(14) 82 − 19 = 63	(15) 68 + 11 = 79
(16) 97 − 38 = 59	(17) 39 + 17 = 56	(18) 73 − 24 = 49	(19) 51 + 14 = 65	(20) 89 − 33 = 56
(21) 26 + 38 = 64	(22) 94 − 16 = 78	(23) 60 + 14 = 74	(24) 87 − 33 = 54	(25) 43 + 18 = 61
(26) 77 − 34 = 43	(27) 31 + 28 = 59	(28) 98 − 46 = 52	(29) 63 + 27 = 90	(30) 92 − 15 = 77

Day 71

(1) 47 + 26 = 73	(2) 62 − 35 = 27	(3) 18 + 79 = 97	(4) 81 − 43 = 38	(5) 15 + 39 = 54
(6) 94 − 58 = 36	(7) 23 + 46 = 69	(8) 75 − 29 = 46	(9) 34 + 62 = 96	(10) 98 − 57 = 41
(11) 28 + 59 = 87	(12) 83 − 47 = 36	(13) 12 + 67 = 79	(14) 88 − 49 = 39	(15) 36 + 42 = 78
(16) 74 − 21 = 53	(17) 57 + 18 = 75	(18) 92 − 34 = 58	(19) 21 + 33 = 54	(20) 63 − 17 = 46
(21) 45 + 32 = 77	(22) 69 − 28 = 41	(23) 27 + 64 = 91	(24) 87 − 39 = 48	(25) 38 + 49 = 87
(26) 72 − 27 = 45	(27) 31 + 56 = 87	(28) 84 − 38 = 46	(29) 19 + 58 = 77	(30) 96 − 43 = 53

Day 72

(1) 53 + 26 = 79	(2) 77 − 18 = 59	(3) 24 + 31 = 55	(4) 66 − 35 = 31	(5) 39 + 23 = 62
(6) 91 − 54 = 37	(7) 17 + 42 = 59	(8) 83 − 38 = 45	(9) 61 + 28 = 89	(10) 95 − 54 = 41
(11) 47 + 21 = 68	(12) 72 − 31 = 41	(13) 29 + 49 = 78	(14) 86 − 38 = 48	(15) 37 + 24 = 61
(16) 79 − 47 = 32	(17) 26 + 68 = 94	(18) 94 − 32 = 62	(19) 31 + 47 = 78	(20) 57 − 22 = 35
(21) 56 + 23 = 79	(22) 77 − 12 = 65	(23) 19 + 34 = 53	(24) 68 − 24 = 44	(25) 52 + 21 = 73
(26) 86 − 33 = 53	(27) 36 + 49 = 85	(28) 91 − 23 = 68	(29) 13 + 48 = 61	(30) 74 − 16 = 58

Solution

Day 73

1) 26 + 41 = 67	2) 83 − 27 = 56	3) 57 + 18 = 75	4) 94 − 44 = 50	5) 38 + 29 = 67
6) 62 − 11 = 51	7) 27 + 37 = 64	8) 59 − 15 = 44	9) 81 + 18 = 99	10) 98 − 49 = 49
11) 43 + 24 = 67	12) 77 − 42 = 35	13) 25 + 45 = 70	14) 85 − 13 = 72	15) 39 + 19 = 58
16) 67 − 22 = 45	17) 14 + 58 = 72	18) 82 − 48 = 34	19) 49 + 18 = 67	20) 76 − 23 = 53
21) 41 + 31 = 72	22) 69 − 27 = 42	23) 18 + 36 = 54	24) 97 − 55 = 42	25) 24 + 15 = 39
26) 83 − 26 = 57	27) 54 + 14 = 68	28) 86 − 35 = 51	29) 47 + 23 = 70	30) 72 − 29 = 43

Day 74

1) 32 + 47 = 79	2) 93 − 48 = 45	3) 19 + 28 = 47	4) 77 − 42 = 35	5) 58 + 15 = 73
6) 89 − 27 = 62	7) 35 + 49 = 84	8) 96 − 43 = 53	9) 41 + 24 = 65	10) 85 − 42 = 43
11) 29 + 39 = 68	12) 74 − 13 = 61	13) 16 + 28 = 44	14) 69 − 17 = 52	15) 21 + 38 = 59
16) 88 − 29 = 59	17) 44 + 23 = 67	18) 97 − 52 = 45	19) 25 + 16 = 41	20) 66 − 17 = 49
21) 13 + 49 = 52	22) 92 − 31 = 61	23) 34 + 47 = 81	24) 78 − 15 = 63	25) 39 + 27 = 66
26) 59 − 24 = 35	27) 17 + 37 = 54	28) 83 − 18 = 65	29) 37 + 24 = 61	30) 96 − 46 = 50

Day 75

1) 36 + 19 = 55	2) 71 − 29 = 42	3) 28 + 38 = 66	4) 85 − 37 = 48	5) 16 + 24 = 40
6) 63 − 12 = 51	7) 22 + 45 = 67	8) 94 − 32 = 62	9) 46 + 27 = 73	10) 73 − 21 = 52
11) 37 + 19 = 56	12) 88 − 46 = 42	13) 25 + 18 = 43	14) 59 − 31 = 28	15) 32 + 46 = 78
16) 78 − 24 = 54	17) 14 + 19 = 33	18) 53 − 17 = 36	19) 18 + 37 = 55	20) 87 − 41 = 46
21) 31 + 15 = 46	22) 68 − 26 = 42	23) 24 + 29 = 53	24) 77 − 34 = 43	25) 14 + 18 = 32
26) 63 − 12 = 51	27) 33 + 47 = 80	28) 92 − 35 = 57	29) 38 + 29 = 67	30) 86 − 13 = 73

Day 76

1) 128 + 234 = 362	2) 672 − 318 = 357	3) 385 + 206 = 591	4) 827 − 132 = 695	5) 553 + 122 = 675
6) 947 − 215 = 732	7) 714 + 132 = 846	8) 625 − 189 = 436	9) 328 + 110 = 438	10) 927 − 408 = 519
11) 455 + 166 = 621	12) 679 − 283 = 396	13) 375 + 146 = 521	14) 848 − 325 = 523	15) 597 + 125 = 722
16) 931 − 217 = 714	17) 496 + 131 = 627	18) 825 − 212 = 613	19) 361 + 102 = 463	20) 948 − 532 = 416
21) 529 + 135 = 664	22) 717 − 205 = 512	23) 482 + 234 = 416	24) 826 − 245 = 581	25) 625 + 185 = 810
26) 845 − 224 = 621	27) 389 + 139 = 528	28) 931 − 307 = 624	29) 689 + 115 = 804	30) 824 − 199 = 625

Solution

Day 77

(1) 518 + 187 = 705	(2) 712 − 203 = 509	(3) 361 + 142 = 503	(4) 938 − 326 = 612	(5) 485 + 211 = 696
(6) 816 − 224 = 592	(7) 628 + 119 = 747	(8) 932 − 314 = 618	(9) 399 + 186 = 585	(10) 841 − 218 = 623
(11) 545 + 126 = 671	(12) 948 − 432 = 516	(13) 628 + 142 = 770	(14) 715 − 206 = 921	(15) 489 + 204 = 693
(16) 913 − 236 = 677	(17) 478 + 155 = 633	(18) 927 − 202 = 725	(19) 555 + 121 = 676	(20) 819 − 244 = 575
(21) 594 + 113 = 707	(22) 826 − 214 = 612	(23) 362 + 133 = 495	(24) 931 − 304 = 627	(25) 618 + 157 = 775
(26) 894 − 378 = 516	(27) 485 + 123 = 608	(28) 716 − 205 = 511	(29) 529 + 147 = 676	(30) 948 − 435 = 513

Day 78

(1) 635 + 210 = 845	(2) 827 − 216 = 611	(3) 268 + 256 = 524	(4) 832 − 220 = 612	(5) 674 + 104 = 778
(6) 915 − 207 = 708	(7) 586 + 155 = 741	(8) 729 − 204 = 525	(9) 387 + 315 = 702	(10) 714 − 115 = 599
(11) 538 + 148 = 686	(12) 648 − 317 = 331	(13) 655 + 310 = 965	(14) 826 − 238 = 588	(15) 398 + 130 = 528
(16) 618 − 248 = 370	(17) 689 + 132 = 821	(18) 628 − 316 = 312	(19) 519 + 138 = 657	(20) 748 − 237 = 511
(21) 485 + 241 = 726	(22) 912 − 304 = 608	(23) 418 + 237 = 655	(24) 825 − 218 = 607	(25) 555 + 115 = 670
(26) 819 − 317 = 502	(27) 679 + 148 = 827	(28) 638 − 314 = 324	(29) 489 + 120 = 609	(30) 726 − 216 = 510

Day 79

(1) 184 + 247 = 431	(2) 572 − 329 = 243	(3) 398 + 214 = 612	(4) 817 − 325 = 492	(5) 419 + 238 = 657
(6) 925 − 308 = 617	(7) 726 + 162 = 888	(8) 856 − 312 = 544	(9) 478 + 183 = 661	(10) 629 − 238 = 391
(11) 341 + 209 = 550	(12) 868 − 293 = 575	(13) 634 + 183 = 817	(14) 626 − 207 = 419	(15) 564 + 218 = 782
(16) 916 − 347 = 569	(17) 693 + 278 = 971	(18) 927 − 279 = 648	(19) 627 + 217 = 844	(20) 895 − 378 = 517
(21) 532 + 163 = 695	(22) 858 − 283 = 575	(23) 479 + 331 = 810	(24) 938 − 379 = 559	(25) 647 + 152 = 799
(26) 978 − 276 = 702	(27) 592 + 216 = 808	(28) 926 − 382 = 544	(29) 541 + 173 = 714	(30) 874 − 238 = 636

Day 80

(1) 625 + 193 = 818	(2) 987 − 527 = 460	(3) 653 + 126 = 779	(4) 812 − 232 = 580	(5) 537 + 214 = 751
(6) 867 − 286 = 581	(7) 654 + 219 = 873	(8) 725 − 318 = 407	(9) 567 + 218 = 785	(10) 924 − 368 = 556
(11) 698 + 214 = 912	(12) 745 − 185 = 560	(13) 417 + 264 = 681	(14) 956 − 628 = 328	(15) 712 + 193 = 905
(16) 829 − 237 = 592	(17) 539 + 317 = 856	(18) 847 − 186 = 661	(19) 524 + 325 = 849	(20) 928 − 472 = 456
(21) 653 + 277 = 930	(22) 775 − 316 = 459	(23) 591 + 217 = 808	(24) 914 − 284 = 630	(25) 326 + 219 = 545
(26) 936 − 327 = 609	(27) 214 + 674 = 888	(28) 828 − 276 = 552	(29) 227 + 514 = 741	(30) 938 − 347 = 591

Solution

Day 81

#	Problem	Answer
1	514 + 236	750
2	827 − 328	499
3	549 + 192	741
4	924 − 327	597
5	654 + 215	869
6	856 − 329	527
7	612 + 107	719
8	928 − 287	641
9	643 + 117	760
10	845 − 175	670
11	569 + 193	762
12	925 − 317	608
13	632 + 219	851
14	917 − 328	589
15	561 + 218	779
16	937 − 289	648
17	647 + 219	866
18	859 − 387	472
19	612 + 218	830
20	937 − 327	610
21	538 + 193	731
22	867 − 278	589
23	628 + 318	946
24	728 − 327	401
25	656 + 317	973
26	875 − 426	449
27	543 + 111	654
28	976 − 238	738
29	615 + 192	807
30	995 − 238	757

Day 82

#	Problem	Answer
1	444 + 319	763
2	938 − 327	611
3	272 + 519	791
4	879 − 589	290
5	263 + 576	839
6	936 − 527	409
7	519 + 338	857
8	649 − 347	302
9	576 + 223	799
10	937 − 227	710
11	541 + 439	980
12	917 − 408	509
13	724 + 185	909
14	666 − 237	429
15	130 + 589	719
16	872 − 236	636
17	734 + 157	891
18	962 − 327	635
19	712 + 136	848
20	927 − 296	631
21	258 + 174	432
22	739 − 235	504
23	576 + 189	765
24	942 − 317	625
25	621 + 238	859
26	897 − 289	608
27	647 + 173	820
28	926 − 307	619
29	679 + 231	910
30	957 − 286	671

Day 83

#	Problem	Answer
1	719 + 158	877
2	864 − 195	669
3	252 + 174	426
4	943 − 217	726
5	596 + 189	785
6	831 − 305	526
7	134 + 757	891
8	869 − 197	672
9	543 + 178	721
10	865 − 216	649
11	638 + 198	836
12	819 − 316	503
13	597 + 136	733
14	847 − 214	633
15	168 + 693	861
16	924 − 259	665
17	657 + 128	785
18	862 − 217	645
19	143 + 767	910
20	828 − 316	512
21	684 + 272	956
22	875 − 338	537
23	659 + 222	881
24	954 − 428	526
25	723 + 116	839
26	932 − 236	696
27	596 + 129	725
28	526 − 118	408
29	192 + 638	830
30	849 − 276	573

Day 84

#	Problem	Answer
1	631 + 249	880
2	956 − 259	697
3	627 + 178	805
4	957 − 328	629
5	689 + 157	846
6	923 − 368	555
7	238 + 617	855
8	845 − 327	518
9	652 + 136	788
10	943 − 317	626
11	189 + 596	785
12	931 − 205	726
13	734 + 158	862
14	765 − 117	648
15	643 + 100	743
16	627 − 166	461
17	136 + 714	850
18	962 − 327	635
19	734 + 157	892
20	872 − 236	636
21	198 + 638	836
22	917 − 316	601
23	136 + 597	733
24	847 − 214	633
25	144 + 693	837
26	924 − 259	665
27	128 + 657	803
28	962 − 327	635
29	167 + 743	910
30	928 − 316	612

Solution

Day 85

#		#		#		#		#	
(1) 684 + 172 = **856**		(2) 875 − 238 = **637**		(3) 659 + 189 = **848**		(4) 854 − 225 = **629**		(5) 723 + 178 = **901**	
(6) 822 − 236 = **586**		(7) 596 + 119 = **715**		(8) 926 − 306 = **620**		(9) 692 + 238 = **930**		(10) 849 − 276 = **573**	
(11) 731 + 159 = **890**		(12) 856 − 259 = **597**		(13) 178 + 627 = **805**		(14) 957 − 328 = **629**		(15) 157 + 689 = **846**	
(16) 923 − 368 = **555**		(17) 238 + 617 = **855**		(18) 845 − 218 = **627**		(19) 136 + 652 = **788**		(20) 943 − 217 = **726**	
(21) 596 + 289 = **885**		(22) 931 − 305 = **626**		(23) 164 + 726 = **890**		(24) 865 − 227 = **638**		(25) 643 + 186 = **829**	
(26) 727 − 196 = **531**		(27) 136 + 712 = **848**		(28) 945 − 228 = **717**		(29) 285 + 418 = **703**		(30) 779 − 489 = **290**	

Day 86

#		#		#		#		#	
(1) 173 + 216 = **389**		(2) 987 − 198 = **789**		(3) 342 + 135 = **447**		(4) 764 − 289 = **475**		(5) 421 + 273 = **694**	
(6) 843 − 287 = **556**		(7) 548 + 146 = **694**		(8) 932 − 418 = **514**		(9) 672 + 182 = **854**		(10) 738 − 482 = **256**	
(11) 563 + 214 = **777**		(12) 827 − 318 = **509**		(13) 714 + 137 = **851**		(14) 852 − 205 = **647**		(15) 624 + 148 = **772**	
(16) 919 − 318 = **601**		(17) 567 + 218 = **785**		(18) 941 − 267 = **674**		(19) 694 + 173 = **867**		(20) 869 − 248 = **621**	
(21) 698 + 231 = **929**		(22) 957 − 369 = **588**		(23) 748 + 172 = **920**		(24) 936 − 226 = **710**		(25) 619 + 245 = **864**	
(26) 928 − 298 = **630**		(27) 675 + 166 = **841**		(28) 894 − 287 = **607**		(29) 541 + 198 = **739**		(30) 978 − 377 = **601**	

Day 87

#		#		#		#		#	
(1) 732 + 168 = **900**		(2) 816 − 207 = **609**		(3) 589 + 194 = **783**		(4) 943 − 386 = **557**		(5) 625 + 269 = **894**	
(6) 749 − 486 = **263**		(7) 186 + 543 = **729**		(8) 917 − 298 = **619**		(9) 687 + 162 = **849**		(10) 932 − 249 = **683**	
(11) 143 + 654 = **797**		(12) 739 − 227 = **512**		(13) 184 + 612 = **769**		(14) 627 − 274 = **353**		(15) 594 + 157 = **751**	
(16) 876 − 217 = **659**		(17) 219 + 675 = **894**		(18) 864 − 345 = **519**		(19) 197 + 541 = **738**		(20) 978 − 327 = **651**	
(21) 732 + 148 = **880**		(22) 816 − 289 = **527**		(23) 589 + 153 = **742**		(24) 943 − 256 = **687**		(25) 625 + 179 = **804**	
(26) 849 − 287 = **562**		(27) 543 + 276 = **819**		(28) 917 − 298 = **619**		(29) 152 + 687 = **839**		(30) 932 − 217 = **715**	

Day 88

#		#		#		#		#	
(1) 256 + 421 = **677**		(2) 832 − 267 = **565**		(3) 473 + 362 = **835**		(4) 745 − 278 = **467**		(5) 534 + 146 = **680**	
(6) 926 − 185 = **741**		(7) 231 + 512 = **743**		(8) 789 − 312 = **477**		(9) 247 + 623 = **870**		(10) 528 − 148 = **380**	
(11) 436 + 214 = **650**		(12) 857 − 269 = **588**		(13) 347 + 521 = **868**		(14) 783 − 124 = **659**		(15) 192 + 428 = **620**	
(16) 734 − 315 = **419**		(17) 327 + 592 = **919**		(18) 875 − 543 = **332**		(19) 178 + 461 = **639**		(20) 845 − 417 = **428**	
(21) 548 + 345 = **893**		(22) 756 − 382 = **374**		(23) 391 + 286 = **477**		(24) 872 − 218 = **654**		(25) 354 + 564 = **918**	
(26) 931 − 482 = **449**		(27) 652 + 238 = **890**		(28) 785 − 356 = **429**		(29) 287 + 498 = **785**		(30) 627 − 146 = **481**	

Solution

Day 89

#	Problem	Answer
1	572 + 314	886
2	816 − 285	531
3	427 + 263	690
4	674 − 295	379
5	215 + 542	757
6	821 − 387	434
7	437 + 259	696
8	674 − 235	439
9	576 + 316	892
10	835 − 417	418
11	621 + 271	892
12	824 − 385	439
13	537 + 316	853
14	826 − 384	442
15	659 + 243	902
16	745 − 356	389
17	564 + 319	883
18	748 − 423	325
19	228 + 645	873
20	915 − 384	531
21	672 + 314	986
22	872 − 437	435
23	596 + 317	913
24	927 − 498	429
25	713 + 194	907
26	945 − 527	418
27	627 + 281	908
28	835 − 468	367
29	529 + 331	860
30	876 − 456	420

Day 90

#	Problem	Answer
1	634 + 268	902
2	786 − 352	434
3	514 + 227	741
4	756 − 362	394
5	453 + 251	704
6	693 − 232	461
7	367 + 245	612
8	751 − 315	436
9	439 + 214	653
10	825 − 437	388
11	586 + 315	901
12	715 − 326	389
13	428 + 169	597
14	765 − 382	383
15	591 + 238	829
16	837 − 398	439
17	448 + 241	689
18	916 − 568	348
19	618 + 194	812
20	845 − 367	478
21	543 + 271	814
22	815 − 372	443
23	497 + 231	728
24	738 − 341	397
25	582 + 198	780
26	927 − 568	359
27	633 + 245	878
28	785 − 358	427
29	319 + 236	555
30	738 − 384	354

Day 91

#	Problem	Answer
1	452 + 218	670
2	692 − 374	318
3	715 + 163	878
4	817 − 436	381
5	527 + 192	719
6	839 − 498	341
7	582 + 271	853
8	928 − 485	443
9	613 + 289	902
10	856 − 364	492
11	567 + 227	340
12	742 − 317	425
13	486 + 216	702
14	715 − 396	319
15	534 + 431	865
16	736 − 487	249
17	623 + 171	794
18	927 − 416	511
19	574 + 319	893
20	819 − 436	383
21	642 + 286	928
22	758 − 369	389
23	573 + 218	791
24	926 − 458	468
25	649 + 315	964
26	835 − 498	337
27	539 + 271	810
28	839 − 549	290
29	638 + 320	958
30	847 − 472	375

Day 92

#	Problem	Answer
1	674 + 317	991
2	846 − 427	419
3	532 + 271	803
4	736 − 368	368
5	458 + 214	672
6	692 − 378	314
7	537 + 129	666
8	726 − 398	328
9	548 + 271	819
10	925 − 458	467
11	657 + 229	886
12	828 − 498	330
13	617 + 271	888
14	916 − 458	458
15	639 + 338	977
16	825 − 498	327
17	271 + 512	783
18	817 − 358	459
19	315 + 534	849
20	738 − 398	340
21	271 + 549	820
22	827 − 498	329
23	239 + 561	800
24	839 − 547	292
25	638 + 271	909
26	654 − 138	516
27	423 + 269	692
28	928 − 357	571
29	543 + 239	782
30	856 − 237	619

Solution

Day 93

#	Problem	Answer
1	372 + 286	658
2	917 − 182	735
3	548 + 327	875
4	872 − 283	589
5	429 + 274	703
6	715 − 286	429
7	321 + 457	778
8	738 − 249	489
9	593 + 316	909
10	832 − 387	445
11	421 + 358	779
12	785 − 246	539
13	632 + 217	849
14	945 − 269	676
15	428 + 417	845
16	864 − 348	516
17	687 + 169	856
18	935 − 327	608
19	547 + 218	765
20	914 − 386	528
21	312 + 287	599
22	858 − 327	531
23	642 + 186	828
24	928 − 317	611
25	569 + 138	707
26	786 − 327	459
27	159 + 623	782
28	917 − 329	588
29	451 + 158	609
30	969 − 328	641

Day 94

#	Problem	Answer
1	456 + 234	690
2	899 − 485	414
3	312 + 578	890
4	843 − 219	624
5	543 + 342	885
6	766 − 347	419
7	654 + 126	780
8	937 − 343	594
9	288 + 519	807
10	728 − 268	460
11	523 + 467	990
12	817 − 235	582
13	365 + 274	639
14	678 − 435	243
15	735 + 189	924
16	957 − 387	570
17	244 + 367	611
18	833 − 298	535
19	441 + 147	588
20	972 − 314	658
21	389 + 512	901
22	757 − 245	512
23	459 + 154	613
24	691 − 219	472
25	537 + 243	780
26	855 − 319	536
27	683 + 164	847
28	927 − 245	682
29	366 + 531	897
30	768 − 285	483

Day 95

#	Problem	Answer
1	488 + 249	737
2	812 − 224	588
3	332 + 477	809
4	685 − 265	420
5	412 + 329	741
6	819 − 245	574
7	288 + 468	756
8	754 − 245	509
9	527 + 198	725
10	861 − 298	563
11	355 + 517	872
12	778 − 266	512
13	475 + 154	629
14	843 − 328	515
15	466 + 199	665
16	912 − 354	558
17	534 + 244	778
18	756 − 219	537
19	288 + 312	600
20	614 − 276	338
21	236 + 366	602
22	819 − 357	462
23	476 + 374	850
24	768 − 289	479
25	389 + 278	667
26	847 − 375	472
27	576 + 184	760
28	836 − 357	479
29	288 + 287	575
30	729 − 394	335

Day 96

#	Problem	Answer
1	457 + 264	721
2	861 − 289	572
3	244 + 167	411
4	799 − 214	555
5	315 + 594	909
6	643 − 255	388
7	376 + 485	861
8	912 − 446	466
9	578 + 395	973
10	747 − 189	558
11	285 + 288	573
12	679 − 398	281
13	511 + 269	780
14	826 − 484	342
15	357 + 279	636
16	683 − 267	416
17	487 + 178	665
18	917 − 415	502
19	366 + 324	690
20	736 − 244	492
21	399 + 278	677
22	827 − 349	478
23	377 + 245	622
24	835 − 367	468
25	355 + 326	681
26	666 − 357	309
27	366 + 489	855
28	727 − 356	371
29	355 + 129	484
30	578 − 245	333

Solution

Day 97

#	Problem	Answer
1	347 + 413	760
2	963 − 475	488
3	412 + 189	601
4	746 − 125	621
5	543 + 284	827
6	917 − 384	533
7	698 + 242	940
8	834 − 543	291
9	323 + 433	756
10	775 − 187	588
11	475 + 359	834
12	812 − 254	558
13	586 + 235	821
14	916 − 325	591
15	329 + 465	794
16	437 − 314	123
17	527 + 294	821
18	765 − 526	239
19	467 + 429	896
20	754 − 346	408
21	498 + 182	680
22	669 − 235	434
23	529 + 348	877
24	964 − 427	537
25	641 + 198	839
26	925 − 316	609
27	547 + 352	899
28	897 − 483	414
29	574 + 289	863
30	843 − 342	501

Day 98

#	Problem	Answer
1	464 + 458	922
2	748 − 285	463
3	587 + 229	816
4	841 − 359	482
5	612 + 377	989
6	946 − 415	531
7	513 + 385	898
8	728 − 274	704
9	632 + 259	891
10	866 − 389	477
11	498 + 283	781
12	731 − 369	362
13	652 + 324	976
14	857 − 461	396
15	365 + 329	694
16	814 − 474	340
17	446 + 355	801
18	976 − 463	513
19	587 + 254	841
20	843 − 424	419
21	587 + 236	823
22	827 − 160	667
23	218 + 457	675
24	861 − 283	578
25	647 + 248	895
26	777 − 486	291
27	429 + 268	697
28	725 − 317	408
29	435 + 357	792
30	654 − 448	206

Day 99

#	Problem	Answer
1	467 + 285	752
2	839 − 358	481
3	426 + 354	780
4	647 − 428	219
5	369 + 249	618
6	754 − 335	419
7	497 + 204	701
8	835 − 329	506
9	329 + 349	678
10	964 − 427	537
11	578 + 136	442
12	836 − 229	607
13	513 + 248	761
14	729 − 285	444
15	435 + 498	933
16	983 − 264	719
17	543 + 257	800
18	835 − 487	348
19	456 + 389	845
20	756 − 318	438
21	169 + 755	924
22	956 − 629	327
23	529 + 238	767
24	795 − 398	397
25	618 + 287	905
26	721 − 315	406
27	534 + 318	852
28	865 − 575	290
29	549 + 285	834
30	723 − 394	329

Day 100

#	Problem	Answer
1	245 + 322	567
2	988 − 129	859
3	487 + 423	910
4	887 − 458	429
5	319 + 594	913
6	955 − 292	663
7	548 + 133	681
8	967 − 343	624
9	427 + 248	675
10	726 − 219	507
11	567 + 159	726
12	989 − 798	191
13	329 + 625	954
14	854 − 488	366
15	636 + 282	918
16	927 − 426	501
17	483 + 318	801
18	955 − 538	417
19	192 + 463	655
20	867 − 539	328
21	137 + 563	700
22	868 − 439	429
23	594 + 315	909
24	976 − 542	434
25	423 + 385	808
26	825 − 362	463
27	528 + 268	796
28	935 − 512	423
29	654 + 186	840
30	845 − 529	316